Student Workbook

for use with

Welding

Principles and Practices

Fifth Edition

Edward R. Bohnart

McGraw
Graw
Hill
Education

Student Workbook for use with

WELDING: PRINCIPLES AND PRACTICES, FIFTH EDITION

Published by McGraw-Hill Education, 2 Penn Plaza, New York, NY 10121. Copyright © 2018 by McGraw-Hill Education. All rights reserved. Printed in the United States of America. Previous editions © 1981, 2005, and 2012. No part of this publication may be reproduced or distributed in any form or by any means, or stored in a database or retrieval system, without the prior written consent of McGraw-Hill Education, including, but not limited to, in any network or other electronic storage or transmission, or broadcast for distance learning.

Some ancillaries, including electronic and print components, may not be available to customers outside the United States.

This book is printed on acid-free paper.

1 2 3 4 5 6   QVS   21 20 19 18 17

ISBN 978-1-259-86989-1
MHID 1-259-86989-X

The Internet addresses listed in the text were accurate at the time of publication. The inclusion of a website does not indicate an endorsement by the authors or McGraw-Hill Education, and McGraw-Hill Education does not guarantee the accuracy of the information presented at these sites.

mheducation.com/highered

Table of Contents

Introduction to the Instructor iv

About the Author iv

Introduction to the Student v

Chapter 1 History of Welding 1

Chapter 2 Industrial Welding 3

Chapter 3 Steel and Other Metals 7

Chapter 4 Basic Joints and Welds 13

Chapter 5 Gas Welding 17

Chapter 6 Flame Cutting Principles 21

Chapter 7 Flame Cutting Practice: Jobs 7-J1–J3 23

Chapter 8 Gas Welding Practice: Jobs 8-J1–J38 27

Chapter 9 Braze Welding and Advanced Gas Welding Practice: Jobs 9-J39–J49 31

Chapter 10 Soldering and Brazing Principles and Practices: Jobs 10-J50–J51 35

Chapter 11 Shielded Metal Arc Welding Principles 39

Chapter 12 Shielded Metal Arc Welding Electrodes 43

Chapter 13 Shielded Metal Arc Welding Practice: Jobs 13-J1–J25 (Plate) 47

Chapter 14 Shielded Metal Arc Welding Practice: Jobs 14-J26–J42 (Plate) 55

Chapter 15 Shielded Metal Arc Welding Practice: Jobs 15-J43–J55 (Plate) 61

Chapter 16 Pipe Welding and Shielded Metal Arc Welding Practice: Jobs 16-J1–J17 (Pipe) 67

Chapter 17 Arc Cutting Principles and Arc Cutting Practice: Jobs 17-J1–J7 73

Chapter 18 Gas Tungsten Arc and Plasma Arc Welding Principles 77

Chapter 19 Gas Tungsten Arc Welding Practice: Jobs 19-J1–J19 (Plate) 81

Chapter 20 Gas Tungsten Arc Welding Practice: Jobs 20-J1–J17 (Pipe) 87

Chapter 21 Gas Metal Arc and Flux Cored Arc Welding Principles 89

Chapter 22 Gas Metal Arc Welding Practice with Solid and Metal Core Wire: Jobs 22-J1–J23 (Plate) 93

Chapter 23 Flux Cored Arc Welding Practice (Plate), Submerged Arc Welding, and Related Processes: FCAW-G Jobs 23-J1–J11, FCAW-S Jobs 23-J1–J12, SAW Job 23-J1 97

Chapter 24 Gas Metal Arc Welding Practice: Jobs 24-J1–J15 (Pipe) 103

Chapter 25 High Energy Beams and Related Welding and Cutting Process Principles 105

Chapter 26 General Equipment for Welding Shops 107

Chapter 27 Automatic and Robotic Arc Welding Equipment 111

Chapter 28 Joint Design, Testing, and Inspection 113

Chapter 29 Reading Shop Drawings 119

Chapter 30 Welding Symbols 123

Chapter 31 Welding and Bonding of Plastics 127

Chapter 32 Safety 131

Appendix A Job Outlines 135

Introduction to the Instructor

This workbook is designed for use with the textbook *Welding: Principles and Practices,* 5/e. Each exercise has been prepared to assist the student in mastering the information in the text. Every chapter in the workbook has the same number as the textbook chapter it reviews.

Every chapter of the text may not be assigned, and chapters need not be covered in sequence. The workbook items are designed to accommodate whatever approach an instructor may decide. The review questions in this workbook are supplementary questions. Also, instructors may wish to develop their own supplementary questions.

A number of approaches are suggested in the assignment of the workbook exercises.

1. The exercises may be completed as "closed book" or "open book" assignments.
2. The material in the textbook may be assigned as homework and the workbook exercises completed at home.
3. The material in the textbook may be assigned as homework and the workbook exercises completed in class.
4. Each student may study the material in the textbook at his or her own rate of speed and complete the workbook exercises on the same basis.

About the Author

Edward R. Bohnart (AWS-SCWI, and former CWE, CWS, CWSR, CRAW-T, and AWS Certified Welder) is the principal of Welding Education and Consulting located in Wisconsin. He launched his consulting business after a successful career with Miller Electric Manufacturing Company, where he directed training operations. He is a graduate of the Nebraska Vocational Technical College in welding and metallurgy and has studied at both the University of Nebraska at Omaha and the University of Omaha.

Bohnart was selected in the 2011 Class of Counselors of the American Welding Society, and he is also an AWS Distinguished Member and national past President. He remains active with the SkillsUSA organization and is past chair of the AWS Skills Competition Committee, which conducted the USOpen Weld Trials to select the TeamUSA welder for the WorldSkills Competition. He was the United States of America's Welding Technical Expert for the WorldSkills Competition from 1989 to 2009. Bohnart chaired the AWS C5 Committee on Arc Welding and Cutting Processes and remains on the committee as an advisor.

The American Welding Society has recognized Ed Bohnart with the National Meritorious Award, George E. Willis Award, and Plummer Memorial Educational Lecture Award. The Wisconsin State Superintendents Technology Education Advisory Committee has acknowledged him with the Technology Literacy Award. The state of Nebraska Community College System has appointed him Alumnus of the Year, and the Youth Development Foundation of SkillsUSA has honored Bohnart with the SkillsUSA Torch Carrier Award.

Ed has been active on the Edison Welding Institute Board of Trustees and on the American Institute of Steel Constructions Industry Round Table. He has served on the industrial advisory boards for Arizona State University, The University of Wisconsin–Stout, Fox Valley Technical Colleges, and the Haney Technical Center industrial advisory boards.

He has lectured at a number of major institutions, such as the Massachusetts Institute of Technology, Colorado School of Mines, Texas A&M, Arizona State University, and the Paton Institute of Welding, Kiev, Ukraine.

Introduction to the Student

Welding is one of the most interesting, challenging, and highest paying jobs in the industry. It also serves as an important tool in the repair and maintenance field. This workbook includes information on all the forms of welding presented in the textbook, *Welding: Principles and Practices,* 5/e. The exercises are intended to help you achieve as complete an understanding of the fundamentals of welding processes and equipment as possible. The questions cover the information a welder must possess in order to become a skilled craftsman and to gain a place of employment in the welding and advanced fields, such as welding inspector, instructor, and foreman.

   The workbook exercises will help you in a number of ways.

1. Surveying the questions will help you determine the information in each chapter to which you should pay special attention.
2. Completing the different kinds of objective questions will be a positive aid in remembering the important aspects of successful welding.
3. Evaluating your success in completing the exercises will indicate not only what you have learned, but also what you need to review.

   The following are examples of the types of questions used in the workbook.

1. *True or False.* Read the statement carefully and decide if it is true or false. Circle *True* if the statement is true and *False* if it is false.

   SAMPLE 1.   Welding is an industry that is rapidly going out of existence. (True or False)

2. *Multiple choice with one right answer.* These items are unfinished sentences with four or five possible endings, only one of which is correct. Read the statement and the possible conclusion. Circle the best answer and/or place its letter in the answer space.

   SAMPLE 2.   St Louis is on the_____c_____.

   a. Rio Grande   **b.** Thames   (**c.** Mississippi)   **d.** Rhine

3. *Multiple choice with one wrong answer.* These items are incomplete statements with four or five possible conclusions, only one of which is wrong. Read the statement and the possible endings. Circle the one that is wrong and/or place its letter in the answer space.

   SAMPLE 3.   All of the following are fractions except_____c_____.

   a. ⅔   **b.** ⅝   (**c.** 6)   **d.** ⅞   **e.** 9/10

4. *Completion.* Completion items are sentences with blanks where key words should be. Study the sentence carefully and determine which word best completes it. Record this word in the answer space.

   SAMPLE 4.   Arc welding is used to join_____metals_____.

5. *Matching.* Matching items are made up of two columns of words or phases. Those on the left are numbered; those on the right are lettered. Determine the numbered item to which the lettered item is best matched, and record the number in the appropriately lettered answer space to the right of the question.

SAMPLE 5.

1. River	a. Everest	a.	2
2. Mountain	b. Atlantic	b.	4
3. Lake	c. Thames	c.	1
4. Ocean	d. Ontario	d.	3

Matching items may also call for matching parts of an illusion to corresponding identifying terms. Answers are recorded in the same was as they are for sample question 5.

6. *Identification.* These questions are usually accompanied by illustrations in which names for pieces of equipment or parts of a joint have been replaced by letters. The correct name for a given article should be recorded in the appropriate lettered answer blank.

SAMPLE 6.   Identify the utensils in Sample Fig 1.1

a.   Knife

b.   Fork

c.   Spoon

Sample Fig. 1-1.

7. *Logical sequence.* Given several steps needed to complete an activity, place them in the order in which they would be undertaken.

Name: _____ Date: _____

CHAPTER 1
History of Welding

Please answer the following questions by choosing the letter of the correct answer, circling true or false, or filling in the blanks.

1. It is always necessary to apply pressure when joining two pieces of metal with the welding process. (True or False)

2. The melting point of filler metals must be higher than 800°F. (True or False)

3. The first metal worked by primitive people was _____.
   a. iron   **b.** bronze   **c.** copper   **d.** aluminum   **e.** silver

4. The first welding took place about _____ years ago.

5. The first experiments from which our present welding processes developed began in the early _____.
   a. 1600s   **b.** 1700   **c.** 1800s   **d.** 1900s

6. Welding was an important industrial process long before World War I. (True or False)

7. The production demands of World War I introduced welding as a means of product _____.
   a. maintenance   **b.** fabrication   **c.** repair   **d.** stabilization   **e.** design

8. The first arc welding was performed with a.c. welding machines. (True or False)

9. The advantages of a.c. welding machines are the absence of arc blow and the high rate of metal deposition. (True or False)

10. The production demands of World War II hastened the development of _____.
    a. bare electrode welding   **b.** inert gas welding
    c. argon/helium welding   **d.** metal-arc weld

11. Early in the twentieth century, _____ was involved in developing the assembly line method for manufacturing automobiles.

12. The development of x-ray of goods called for improved methods of fabrication. Examination of weld metal made it possible to examine the internal soundness of welded joints. (True or False)

13. The drawbacks of inert gas welding were the cost of the _____ and the lack of suitable equipment.

14. _____ welding was first used to weld aluminum and magnesium.

15. MIG/MAG welding was patented before TIG welding. (True or False)

16. MIG/MAG welding has three of the following advantages. Which one is incorrect?
   a. small heat affected zone   b. wide bead width
   c. deep penetration   d. faster welding speeds

17. Which of these processes is NOT an arc welding process?
   a. atomic-hydrogen welding   b. plasma arc welding
   c. stud welding   d. forge welding
   e. electrogas welding

18. Today there are about 25 welding processes in use. (True or False)

19. Welding is done in every civilized country in the world. (True or False)

20. Women cannot be welders. (True or False)

21. Welding has had very little development in other countries of the world. (True or False)

22. Which of the following is NOT considered a welding process?
   a. gas   b. resistance   c. brazing   d. soldering   e. arc

23. Many qualified welders are certified under these welding codes: (which does NOT apply)
   a. AWS   b. ASME   c. ADI   d. API

24. A diploma from a welding training program is all the certification the bearer needs to perform any welding job. (True or False)

25. A welder is rarely called upon to do overhead welding. (True or False)

26. Lives may depend upon the quality of a welder's work. (True or False)

27. A welder must be able to weld metals other than steel. (True or False)

28. In the welding field, there is little opportunity for advancement beyond the job of welder. (True or False)

29. In doing a job, each welder takes precautionary measures for only his own safety. (True or False)

30. Safety precautions should be followed both in the school shop and in industry. (True or False)

Name: _____ Date: _____

CHAPTER 2
Industrial Welding

Please answer the following questions by choosing the letter of the correct answer, circling true or false, or filling in the blanks.

1. The industrial functions of welding are _____ maintenance, and repair.

2. The use of rolled steel in fabrication has the following advantages over the use of castings except one. Which is incorrect?

    a. stronger
    b. stiffer
    c. more uniform
    d. cheaper
    e. rust free

3. High stresses, high temperatures, and high speeds make it difficult to use welding extensively in the space industry. (True or False)

4. Which of the following is not an advantage of welded construction equipment?

    a. economy
    b. size limitation
    c. strength
    d. rigidity
    e. lightweight

5. The welded joint can be substituted for heavy reinforcing sections in the fabrication of construction equipment. (True or False)

6. The welded joint is not quite as strong as the sections it joins. (True or False)

7. Robotic welding is limited to the aerospace industry. (True or False)

8. Since the mid 1990's, America's road building/repair program has diminished. (True or False)

9. When fully loaded, a giant off-highway truck can hold _____ cubic yards of earth.

10. The following welding processes play an important part in household equipment. Which one does not?

    a. GTAW
    b. GMAW
    c. SAW
    d. brazing

11. Although steel has many good qualities, cast iron is still the better material for use in machine tools. (True or False)

12. The production of nuclear energy would not be possible without the use of the welding process. (True or False)

13. Welded pipe is well suited for use in nuclear plants because it can withstand the high heat and _____ required to operate such installations.

14. Welded piping systems reduce maintenance costs because their permanently tight connections have greater _____ and rigidity.

15. Welded fittings make it possible to fabricate any combination of sizes and shapes. (True or False)

16. The use of welding on overland pipelines is limited by their extreme length. (True or False)

17. Although welding allows more flexibility in the design of freight cars, it also increases a car's weight. (True or False)

18. A relatively new development in freight-car construction is the super-size, all-welded, aluminum _____.

19. Three manual welding processes are used in the fabrication of railroad equipment. Which of the following is not?
    a. shielded metal-arc
    b. TIG
    c. plasma arc
    d. MIG/MAG

20. Welding is the principal method of joining materials used by the railroad industry. (True or False)

21. Welded passenger coaches are stronger, _____, and more comfortable than riveted coaches.

22. Which one of the following four is not an advantage of welded submarines?
    a. reduced size
    b. stronger hulls
    c. greater resistance to depth bombs
    d. caulked seams

23. The welded seams of ships' hulls are difficult to repair. (True or False)

24. Prefabrication, preassembly, and welding were first used in the construction of _____ ships during World War II.

25. For over 50 years, bridges have been constructed wholly or in part by the welding process. (True or False)

26. A welded butt joint has less internal stress than a riveted joint. (True or False)

27. In building construction, welding is limited to the structural steel frame work and to piping. (True or False)

28. Steel home construction save owners thousands in upkeep, insurance, and _____.

29. Two serious service failures of riveted tanks were _____ and corrosion around rivets.

30. Welding is used extensively in the fabrication of tanks and pressure vessels because of three of the following. Which one is incorrect?
   a. requires less material
   b. develops strength equal to 90% of the tank plate
   c. eliminates seam caulking
   d. increases fabrication speed

Name: _____ Date: _____

CHAPTER 3

Steel and Other Metals

Please answer the following questions by choosing the letter of the correct answer, circling true or false, or filling in the blanks.

1. Metals are separated into two major groups: ferrous and _____.

2. Metals that have high iron content are designated as _____ metals.

3. Which of the following is not a nonferrous metals?
   a. nickel   **b.** platinum   **c.** steel alloys
   d. radioactive metals   **e.** aluminum

4. Iron is an alloy of many metals. (True or False)

5. Iron ore is found in abundance only in the U.S. and European countries. (True or False)

6. Iron and _____ are combined to make steel.

7. More welding is performed on _____ than on any other metal.

8. The first people to record the use of iron were the _____.

9. The shaft furnaces used in Europe after A.D. 1350 were the predecessors of modern _____ furnaces.

10. The _____ and cementation processes were used to make steel from ancient times until the development of the bessemer process.

11. Of the two older steelmaking processes mentioned in question 10, only the _____ process is still in use today.

12. Steelmaking started in the United States about _____ years ago.

13. Several events spurred the growth of the steel industry. Which of the following did not?
   a. new uses for iron
   b. the shortage of aluminum
   c. the discovery of large iron ore deposits in Michigan
   d. the development of the Bessemer and open hearth process
   e. the expansion of the railroads

14. Steel making facilities have not changed over the last few decades. (True or False)

15. Japan is the world's largest producer of steel. (True or False)

16. The perfection of _____ as a means of joining metals has expanded the use of steel.

17. In the United States, nearly all the ore is mined in northern Minnesota near Lake Superior. (True or False)

18. The purest iron ore comes from (one right):
    a. United States   **b.** India   **c.** Brazil   **d.** Sweden   **e.** Liberia

19. Iron ore straight from the mine is of suitable quality to feed the blast furnace. (True or False)

20. Oxygen is the most abundant element on earth. (True or False)

21. Oxygen is used in steelmaking to _____ the product and speed up the process.

22. Fuels burned in pure oxygen produce much _____ temperatures than fuels burned in air.

23. Heat energy for steelmaking comes from four fuels, three of which are found in nature. Which of the following does not belong?
    a. coal   **b.** oil   **c.** coke   **d.** natural gas

24. Of the four fuels mentioned in question 23, _____ is the most important to the iron and steel industry.

25. _____ is made by heating coal to a high temperature in the absence of air.

26. Nearly 66% of the steel being currently used is recycled. (True or False)

27. Scrap steel has become such a valuable commodity that the American Metal Market tracks the price of certain grades of scrap daily. (True or False)

28. _____ is used as a flux in the blast furnace.

29. Fluxes are used in steelmaking to separate the _____ from the iron ore.

30. _____ materials do not melt easily.

31. _____ is the residue produced by steelmaking.

32. Pure carbon exists in two crystalline forms: diamonds and _____.

33. In the _____ furnace, iron is separated from most of the impurities in the ore.

34. For making steel, the blast furnace is charged with three of the following. Which is incorrect?
   a. limestone   **b.** coal   **c.** iron ore   **d.** coke

35. Liquid iron is poured into molds to make _____ iron.

36. The annual production of the product referred to in question 35 has declined because the number of blast furnaces is decreasing. (True or False)

37. Melting steel in a _____ purifies the finished product by reducing the amount of unwanted gases in the metal.

38. The two processes for vacuum melting are vacuum _____ melting and consumable electrode vacuum arc melting.

39. Match the methods of degassing steel to their best descriptions.
   1. Stream   **a.** Air is removed from a tank containing   **a.** _____
                              molten steel
   2. Ladle   **b.** Molten steel is forced through a nozzle into   **b.** _____
                              a vacuum chamber
   3. Vacuum lifter   **c.** Molten steel is poured into a tank from   **c.** _____
                              which the air has already been removed

40. From the blast furnace, molten steel is poured into _____ molds for cooling.

41. The method of reducing metal to the desired shape is known as _____.

42. After rolling, ingots are known as blooms, billets, or _____, depending on their size and shape.

43. Match the terms for various types of flat-rolled steel to their best descriptions.
   1. Black iron   **a.** Coated with zinc   **a.** _____
   2. Galvanized   **b.** Coated with lead and tin **b.** _____
   3. Tin plate   **c.** Untreated, hot rolled   **c.** _____
   4. Terne plate   **d.** Coated with tin   **d.** _____

44. Tubular steel products are referred to as seamless or _____, depending upon how they are manufactured.

45. An extrusion is formed by drawing metal through an opening. (True or False)

46. When metal is hammered, rolled, or drawn at ordinary temperatures, it is called
_____ working.

47. Heating and cooling a metal to improve its physical or structural properties is called
_____ treatment.

48. Match the following heat treatments to the phrases which best describe them.

1. Hardening	**a.** Alters ductility, toughness, or electrical or magnetic properties	**a.** _____
2. Case hardening	**b.** Produces wear-resistant surface with soft, tough interior	**b.** _____
3. Annealing	**c.** Reduces hardness after heat treatment and relieves stresses caused by quenching	**c.** _____
4. Tempering	**d.** Improves grain structure after welding, casting, or forging	**d.** _____
5. Normalizing	**e.** Increases hardness of medium to very high carbon steel	**e.** _____

49. Match the terms used to describe various physical properties of metals to their definitions.

1. Melting point	**a.** Ease with which a metal is vaporized	**a.** _____
2. Fusibility	**b.** Brittleness of hot metal	**b.** _____
3. Volatility	**c.** Temperature at which a solid becomes liquid	**c.** _____
4. Thermal conductivity	**d.** Property affected when metal contains slag, inclusions, and gas pockets	**d.** _____
5. Hot shortness	**e.** Ability to change shape without breaking	**e.** _____
6. Density	**f.** Ease with which a metal may be melted	**f.** _____
7. Plasticity	**g.** Ability to carry heat	**g.** _____

50. Desirable qualities in metals include all but which of the following?

    a. strength   **b.** toughness   **c.** shock resistance

    d. brittleness   **e.** hardness

51. The science that deals with the internal structure of metals is called
_____.

52. Failure of metals under repeated or alternating stresses is known as _____ failure.

53. _____ loading is another way of referring to fatigue testing.

54. Aluminum has a higher melting point than lead. (True or False)

55. An increase in carbon content in steel makes the steel softer. (True or False)

56. All but one of the following are among the desirable qualities of copper. Which one is incorrect?
    a. malleability   **b.** ductility   **c.** corrosion resistance
    d. hardness   **e.** thermal conductivity

57. Sulfur improves the quality of steel. (True or False)

58. Aluminum is found in abundance in the pure state. (True or False)

59. All but one of the following metals increase the corrosion resistance of steel. Which one is incorrect?
    a. manganese   **b.** molybdenum   **c.** chromium   **d.** nickel   **e.** niobium

60. High carbon steels have a carbon content of approximately 0.30 to 0.60 per cent. (True or False)

61. The metal most commonly added to iron and carbon to improve the corrosion resistance of stainless steel is _____.

62. Tool steels have low carbon content. (True or False)

63. The four types of cast iron are (one wrong):
    a. gray   **b.** malleable   **c.** ductile   **d.** white   **e.** nodular

64. The two major aspects of contraction in all types of welding are distortion and
_____.

65. Steps that can be taken before welding to prevent distortion include all but one of the following. Which one does not belong?
    a. arranging joints so they balance each other
    b. placing parts in the exact position they should occupy when the weld is completed
    c. prebending parts to be welded
    d. using semi-automatic, submerged arc welding processes

66. Welding both sides of a joint at the same time virtually eliminates distortion. (True or False)

67. All metals expand at the same rate. (True or False)

68. Match the methods of controlling residual stress to their best descriptions,

1. Postheating
2. Full annealing
3. Cold peening
4. Preheating

a. Controls expansion and contraction during the welding operation

b. Most effective method, but difficult to control

c. Most commonly used method

d. Stretches bead by hammering

a. _____

b. _____

c. _____

d. _____

69. Using a low frequency, high amplitude vibration to reduce residual stress levels to a point where they cannot cause distortion or other problems is known as _____ stress relieving.

Name: _____ Date: _____

CHAPTER 4
Basic Joints and Welds

Please answer the following questions by choosing the letter of the correct answer, circling true or false, or filling in the blanks.

1. Identify the five basic types of joints shown in Fig. 4-1.

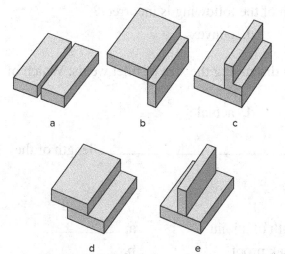

a.
b.
c.
d.
e.

a. _____

b. _____

c. _____

d. _____

e. _____

Fig. 4-1.

2. The welder need not be concerned about design and welding procedures. This is the responsibility of the engineer. (True or False)

3. Which of the following is not one of the four basic types of welds?
   a. bead   **b.** groove   **c.** beam   **d.** fillet   **e.** plug

4. Open corner joints are also welded with fillet welds. (True or False)

5. Groove welds are used for butt joints. (True or False)

6. Strength welds are usually designed to possess the maximum _____ characteristics of the base metal.

7. Strength welds may be all but one of the following. Which one is incorrect?
   a. bead   **b.** edge   **c.** groove   **d.** fillet

8. A groove weld is measured and sized by its depth of _____ fusion into the joint.

9. Steel is classified as alloy steel when the content of alloying elements exceeds certain limits. (True or False)

10. Proper reinforcement should not exceed _____ inch.

11. Which of the following is not part of a ¾ groove weld?

    a. throat   **b.** size   **c.** shoulder   **d.** overlap   **e.** face

12. The most common weld used in industry is the _____ weld.

13. Contour is the shape of the _____ of the weld.

14. Fillet welds can have three types of contours. Which of the following is incorrect?

    a. context   **b.** concave   **c.** flat   **d.** convex

15. Three different throat sizes may be referred to when discussing the size of fillet welds. Which of the following is incorrect?

    a. theoretical   **b.** effective   **c.** acute   **d.** actual

16. Fillet and groove welds are usually made along the _____ length of the joint.

17. Match the types of welds to their best descriptions.

1. Strength	**a.** Extends the entire length of a joint	**a.** _____
2. Caulking	**b.** Makes riveted joints leak proof	**b.** _____
3. Composite	**c.** Holds job parts together in preparation for welding	**c.** _____
4. Continuous	**d.** Carries the structural load	**d.** _____
5. Intermittent	**e.** Placed at intervals to reduce cost of noncritical work	**e.** _____
6. Tack	**f.** Meets load requirements and is also leak proof	**f.** _____

18. Match the welding positions to their descriptions.

1. Flat	**a.** Weld travel may be up or down	**a.** _____
2. Horizontal	**b.** Filler is deposited from upper side of joint; weld face is horizontal	**b.** _____
3. Vertical	**c.** Filler metal is deposited from underside of joint; weld face is horizontal	**c.** _____
4. Overhead	**d.** Filler is deposited on upper side of horizontal surface against a vertical surface	**d.** _____

19. Metal extending above the surfaces of the plate on a groove weld is called

_____.

20. High weld reinforcement increases the strength of a weld. (True or False)

21. The width of a groove weld should not extend beyond the shoulder of the joint more than:

    a. ⅜″   **b.** ½″   **c.** ¹⁄₁₆″   **d.** ⅛″   **e.** ¼″

22. The distance from the root to the toe of a fillet weld is called the _____.

23. Throat thickness is the distance from the root to the _____ of a fillet weld.

24. The ideal fillet weld has a _____ or slightly convex face and equal leg length.

25. A concave fillet weld is stronger and more economical than other fillet welds. (True or False)

26. Generally, welded joints are at least as strong as the base metal being welded. (True or False)

27. The strength of a joint depends upon all but one of the following. Which one is incorrect?

    a. strength of the weld metal

    b. type of joint penetration

    c. heat treatment

    d. position of welding

    e. location of joint in the assembly

28. A weld _____ is any interruption in normal flow of the structure of a weldment.

29. Match the weld defects to their best descriptions.

1. Insufficient throat	**a.** Incomplete fusion in a fillet weld, leaving no root penetration	**a.** _____
2. Excessive convexity	**b.** An overflow of weld metal beyond the end of fusion	**b.** _____
3. Undercut	**c.** Gas entrapped in the filler metal	**c.** _____
4. Overlap	**d.** Fillet weld weakened by shortening a leg	**d.** _____
5. Insufficient leg	**e.** Throat thickness less than specified thickness on a fillet weld	**e.** _____
6. Bridging	**f.** Opposite of a concave profile	**f.** _____
7. Porosity	**g.** A cutting away of plate surface at the weld toe reduces plate thickness	**g.** _____

Name: _____ Date: _____

CHAPTER 5
Gas Welding

Please answer the following questions by choosing the letter of the correct answer, circling true or false, or filling in the blanks.

1. The oxyacetylene process is used for all but one of the following operations. Which one is incorrect?
   - **a.** brazing
   - **b.** soldering
   - **c.** metalizing
   - **d.** large diameter pipe welding
   - **e.** general maintenance and repair

2. The fuel gases commonly used for gas welding, cutting, and heating include all but one of the following. Which one is incorrect?
   - **a.** acetylene
   - **b.** Mapp®
   - **c.** propane
   - **d.** coal gas
   - **e.** natural gas

3. Oxygen burns at a very high temperature. (True or False)

4. Oxygen may be manufactured by the liquefaction of air or the _____ of water.

5. To conserve oxygen, open cylinder valves only part way. (True or False)

6. Oxygen cylinder valves should be greased regularly so that they operate properly. (True or False)

7. Because of the pressure involved, oxygen cylinders undergo rigid testing and inspection. (True or False)

8. Acetylene is generated as the result of a chemical reaction between calcium carbide and _____.

9. Because of the danger of storing highly pressurized acetylene, cylinders are filled with _____.

10. For safety purposes, open acetylene cylinder valves only partway. (True or False)

11. Match the fuel gases to the temperature of their flames under neutral conditions.
    1. Acetylene   **a.** 4,600° F   **a.** _____
    2. Mapp®   **b.** 5,301° F   **b.** _____
    3. Propane   **c.** 5,190° F   **c.** _____
    4. Natural gas   **d.** 5,420° F   **d.** _____
    5. Hydrogen   **e.** 5,000° F   **e.** _____

12. An effective welding fuel gas must possess all but one of the following characteristics. Which one is incorrect?

   a. high flame temperature

   b. high rate of flame propagation

   c. sharp preheat cones

   d. adequate heat

   e. minimum chemical reaction of the flame with base and filler metal

13. An acetylene cylinder may be stored lying on its side. (True or False)

14. Mapp® gas is the safest industrial fuel. (True or False)

15. All portable cylinders used for the storage and shipment of compressed gases shall be constructed and maintained in accordance with the regulations of the _____.

16. Approved protective equipment shall be installed in OAW systems to prevent all but one of the following. Which one is incorrect?

   a. backflow of oxygen into the fuel-gas supply system

   b. over acetylene gas pressure

   c. passage of a flashback into the fuel-gas supply system

   d. excessive back pressure of oxygen in the fuel-gas supply system

17. Increasing the temperature of a cylinder of gas will _____ the pressure within the tank.

18. Match the pieces of equipment to their functions.

1. Torch	a. Ignites the oxyacetylene mixture	a. _____
2. Regulator	b. Carries the oxygen and acetylene cylinders	b. _____
3. Hose	c. Used for cutting	c. _____
4. Goggles	d. Reduces tank pressure to torch pressure	d. _____
5. Lighter	e. Carries gas from tank to torch	e. _____
6. Truck	f. Required for safety	f. _____

19. Single-stage regulators are for use with _____ systems.

20. The two types of torches are the balanced pressure and the _____.

21. Use torch _____ to produce a variety of flame sizes.

22. Most welding tips are made of pure drawn _____.

23. The following gases use which colored hoses?

1. Green		**a.** Acetylene		**a.** _____
2. Red		**b.** Air		**b.** _____
3. Black		**c.** Oxygen		**c.** _____

24. Oxygen-hose couplings have _____ hand threads.

25. A _____ is a cleaning agent used to dissolve oxides and cleanse metals for welding, soldering, and brazing.

26. Hydrogen welding is used primarily for welding metals with _____ melting points.

27. A special regulator is needed when using propane, butane, city gas, or natural gas as a fuel gas for welding. (True or False)

28. _____ gas is used extensively in the plumbing, refrigeration, and electrical trades for heating and soldering operations.

29. Acetylene burned with air produces lower flame temperatures than the other gas combinations. (True or False)

30. Carbon in the form of rods, plates, or paste is used for all but one of the following. Which one is incorrect?

a. to protect surfaces and holes

b. to make back-up welds

c. to supply additional carbon to the molten metal

d. to control and shape the flow of metal

e. to support and align broken parts

31. Eyeglass-type frames are recommended for welding lenses to be worn indoors because they are light and comfortable to wear. (True or False)

32. Match the welding lens shade numbers to the type of work for which they are recommended.

1. Shade 2	**a.** Light brazing		**a.** _____
2. Shade 3	**b.** Standard acetylene gas welding		**b.** _____
3. Shade 4	**c.** Acetylene brazing		**c.** _____
4. Shade 5	**d.** Carbon arc cutting and welding		**d.** _____
5. Shade 6	**e.** Electric arc welding between 75 and 250 amperes		**e.** _____
6. Shade 8	**f.** Low-amperage electric arc welding		**f.** _____
7. Shade 10	**g.** Electric arc cutting above 250 amperes		**g.** _____
8. Shade 12	**h.** Light acetylene cutting		**h.** _____
9. Shade 14	**i.** Low temperature furnace work		**i.** _____

CHAPTER 6
Flame Cutting Principles

Please answer the following questions by choosing the letter of the correct answer, circling true or false, or filling in the blanks.

1. Oxyacetylene cutting is excellent for cutting nonferrous metals. (True or False)

2. Oxyacetylene cutting is a very rapid form of rusting. (True or False)

3. The oxy-fuel gas cutting process uses only acetylene as a fuel gas. (True or False)

4. An adaptable cutting attachment makes it possible to use a welding torch as a cutting tool without disconnecting the hose. (True or False)

5. The central hole in a cutting tip provides for the passage of _____.

6. The holes around the center hole of a cutting tip are _____ flame holes.

7. A cutting torch should be ignited only with a friction lighter or safety matches. (True or False)

8. Both goggles and _____ must be worn during the cutting process for protection from sparks, metal particles, and heat.

9. To cut straight lines accurately, you use which of the following?
    a. straight gouging squares
    b. magnetic burning tips
    c. rivet-burning tips
    d. slicing gauging tips
    e. straight edge or template

10. Oxy-fuel gas cutting machines are able to cut only a limited number of designs. (True or False)

11. Cutting tips are designated as standard or _____.

12. The various fuel gases require different volumes of oxygen and _____.

13. It is possible to cut geometric designs into metal sheets without using templates. (True or False)

14. For plates up to ½-inch thick, _____ cutting is an efficient method of cutting several thicknesses of material at once.

15. A portable _____ cutter may be used for trimming and beveling angles and channels.

16. A length of black iron pipe attached to a oxygen source and used for cutting heavy sections of steel is an oxygen _____ cutter.

17. The beam cutter is a portable structural fabricating tool. From one rail setting the operator can trim, bevel, and cope beams, channels, and angles. (True or False)

Name: _____ Date: _____

CHAPTER 7
Flame Cutting Practice: Jobs 7-J1–J3

Please answer the following questions by choosing the letter of the correct answer, circling true or false, or filling in the blanks.

1. Flame cutting is an electric welding process. (True or False)

2. Flame-cutting processes depend upon the fact that all metals _____ to a certain degree.

3. The cutting torch provides the heating flame, maintains the temperature, and directs a _____ stream on the cutting point.

4. Nearly all flame-cutting problems are caused by obstruction of the nozzle tip. (True or False)

5. Mechanical cutting may damage the plate edge of the cut metal. (True or False)

6. The flame-cut edge of high carbon steel has a tendency to harden and crack. (True or False)

7. The function of the fuel gas is to feed the _____ flames.

8. Of the following fuel gases, which is most commonly used?
   a. propane   **b.** Mapp®   **c.** acetogen   **d.** acetylene   **e.** natural gas

9. The two gases preferred for underwater cutting are _____.
   a. acetylene and acetogen   **b.** natural gas and hydrogen
   c. Mapp® gas and propane   **d.** acetylene and hydrogen

10. The hottest flame is produced by _____.
    a. acetylene   **b.** propane   **c.** natural gas   **d.** hydrogen   **e.** acetogen

11. Mapp® gas produces more heat, measured in Btus, than acetylene. (True or False)

12. The fuel gas that concentrates the most Btus in one area is _____.
    a. acetylene   **b.** propane   **c.** natural gas   **d.** hydrogen   **e.** Mapp® gas

13. Acetylene is expensive to use because it requires more oxygen for the cutting process than other gases do. (True or False)

14. The most economical fuel gas, widely used in steel mills for removing surface defects, is _____.

    a. acetylene   **b.** propane   **c.** natural gas   **d.** hydrogen   **e.** Mapp® gas

15. For cutting plate six or more inches thick, the better choices are Mapp® gas or
_____.

16. Oxyacetylene cutting is most commonly used on _____ and low alloy
steels.

17. The rate of oxidation decreases as the carbon content of metals increases. (True or False)

18. All metal oxides melt at a lower temperature than the base metal. (True or False)

19. Match the methods for speeding the cutting process to their best descriptions.

1. Preheating	**a.** Melting a low carbon steel plate forms an excess of iron oxide, allowing a continuous cut	**a.** _____
2. Waster plate	**b.** Moving the torch from side to side oxidizes additional material	**b.** _____
3. Wire feed	**c.** Increasing the temperature of the material to be cut increases the rate of oxidation	**c.** _____
4. Oscillatory motion	**d.** Burning low carbon steel wire brings the surface of a metal plate to ignition temperature	**d.** _____

20. The gap created as material is removed by cutting is called the _____.

21. The width of the gap produced by the cut increases as the thickness of the material increases. (True or False)

22. The flow of high-pressure oxygen may form _____ lines on the face of the work.

23. Match the types of cutting to their best descriptions.

1. Straight line cutting	**a.** Forms a hole in a metal part	**a.** _____
2. Bevel cutting	**b.** Torch is held perpendicular to plate	**b.** _____
3. Flame piercing	**c.** Removes a narrow strip of surface metal without penetrating the plate	**c.** _____
4. Flame scarfing	**d.** Removes a threaded bolt from a threaded hole without destroying hole or threads	**d.** _____
5. Flame washing	**e.** Removes cracks, surface seams, scabs, and other defects from unfinished steel	**e.** _____
6. Flame gouging	**f.** Torch tip is held sideways at a designated angle	**f.** _____

24. One of the principal uses for oxyacetylene flame cutting is the preparation of plate and
_____ for welded fabrications.

Name: _____ Date: _____

25. Holding the nozzle tip too close to the plate will cause _____ at the top edge.

26. Unsteady torch travel produces a wavy, _____ cut.

27. Preheat flames that are too _____ will cause too much dross.

28. Inadequate preheat with flames held too far from the plate produces a _____ too wide at the top.

29. To increase cutting speed for thicker materials, it is necessary to _____ tip size.

30. Increase the working pressure on the regulators by turning the adjusting screw to the _____.

31. A carburizing flame contains an excessive amount of _____.

32. As the oxygen valve is adjusted so that the secondary cone of the carburizing flame disappears, the _____ flame is formed.

33. If a cut surface is to be used for welding, a _____ flame is recommended.

34. The fastest preheat time is achieved with a reducing flame. (True or False)

35. For a cutting operation, the preheat flames should be in contact with the metal. (True or False)

36. A cut that has been started properly will produce a shower of _____ on the underside of the plate.

37. During a satisfactory cut, dross will flow freely from the _____.

38. The deposit resulting from the oxygen cutting process, which adheres to the base metal or cut surface, is known as _____.

CHAPTER 8

Gas Welding Practice: Jobs 8-J1–J38

Please answer the following questions by choosing the letter of the correct answer, circling true or false, or filling in the blanks.

1. Match the following terms to their best definitions.

   1. Fusion
   2. Fine ripples with even width
   3. Chamfering
   4. Penetration
   5. Reinforcement

   a. Grooving
   b. Building weld metal above the surface of the base metal
   c. Failure of base and filler metals to join completely
   d. Depth to which base metal is melted and joined
   e. The characteristic of a good weld

   a. _____
   b. _____
   c. _____
   d. _____
   e. _____

2. A neutral flame does all but one of the following. Which one is incorrect?

   a. causes no chemical change in the weld metal
   b. produces the highest temperature of the three oxyacetylene flames
   c. serves as the reference point for flame adjustments
   d. is used for most oxyacetylene cutting and welding operations

3. Three of the following four are true concerning an excess acetylene flame. Which one is incorrect?

   a. introduces carbon into the weld pool
   b. is a reducing flame
   c. creates a cloudy, boiling weld pool when used on steel
   d. produces a ductile but tough weld

4. An excess oxygen flame does all but one of the following. Which one is incorrect?

   a. removes excess oxygen from iron oxides when used on steel
   b. reduces flame temperature with too great an oxygen increase
   c. produces a flame hotter than the carburizing flame
   d. seriously reduces weld quality when used with oxidizing metals

5. A carburizing flame is recommended for all but one of the following. Which one is incorrect?

   a. high-test pipe   b. wrought iron   c. chrome   d. alloy steels

6. An oxidizing flame is recommended for all but one of the following. Which one is incorrect?

   a. sheet brass   b. bronze   c. chromium nickel   d. steel plate

7. Maintenance welding requires the use of _____ welding equipment.

8. An oxygen hose may be green or black. (True or False)

9. An acetylene hose may be red or white. (True or False)

10. All right-hand threads are _____ connections.

11. High gas pressure is often the major cause of poor _____.

12. When closing down equipment, turn off the _____ valve on the torch first.

13. When closing down portable or line equipment, always release the regulator adjusting screws at the proper step. (True or False)

14. After the cylinder valves are closed, torch valves are re-opened to drain gas lines. (True or False)

15. Neither check valves nor flashback arrestors will protect the torch or tip. (True or False)

16. Information on safety operation procedures can be obtained by contacting the _____.

17. An important factor in making a ripple weld is the _____ speed of the flame.

18. Moving the welding flame across the plate too fast causes all but one of the following. Which one is incorrect?
    a. loss of weld pool   b. poor fusion
    c. burn-through   d. a narrow bead

19. A satisfactory ripple weld has even ripples of uniform width. (True or False)

20. To form the molten pool for ripple welding, hold the inner core of the torch flame _____ inch above the plate surface.

21. For ripple welding in the vertical position, hold the torch like a pencil. (True or False)

22. Edge and corner joints may be welded with or without filler rod. (True or False)

23. If the filler rod is permitted to drip from above the weld pool, the molten drop will become _____.

24. In bead welding, the cone of the flame should touch the molten weld pool. (True or False)

25. If the filler rod is of the proper size it should not melt in the _____.

26. The _____ welding position is often the most difficult to master because of the weld's tendency to sag.

27. To be sure of maximum strength for welded edge or corner joints, filler rod must be used. (True or False)

28. Corner-joint penetration should show through to the backside of the weld. (True or False)

29. Use the torch flame to protect the weld pool from the formation of oxides and _____.

30. Square butt joints require no edge preparation for welding. (True or False)

31. In making a butt weld, melt the plate _____ before adding filler rod.

32. Groove-weld reinforcement should be no higher than.
    a. ⅟₃₂ inch   **b.** ³⁄₃₂ inch   **c.** ⅟₁₆ inch   **d.** ⅛ inch

33. Bending a good butt weld will produce no cracks or breaks. (True or False)

34. A lap joint is welded with a _____ weld.

35. Maximum lap-joint strength requires welding on both sides. (True or False)

36. It is possible for the top sheet to burn away during lap welding while the bottom sheet remains too cool. (True or False)

37. During lap welding, protect the top plate from excess heat with which one of the following?
    a. filler rod   **b.** torch position   **c.** the weld pool   **d.** tack welds

38. The difference between sheet steel and steel plate is the _____ of the metal.

39. When welding a T-joint, all but one of the following is correct. Which one is incorrect?
    a. the vertical plate tends to melt first
    b. use the filler rod to help prevent undercut
    c. use less heat than you would for other joints
    d. form a slightly convex weld face

40. When welding the second side of a T-joint, direct more heat toward the flat plate. (True or False)

41. Light gauge sheet metal should be welded with a _____ type-welding torch.

42. Pipe and heavy plate are welded with _____ test filler rod.

43. For _____ plate welding, the travel is from left to right and the flame is directed backward.

44. The surface appearance of a backhand weld is smoother than that of a forehand weld. (True or False)

45. Because of the constantly changing torch and rod angles required, _____ welding is recommended for a welder's first pipe practice.

46. For pipe welding, use one weld pass for each ⅛-inch thickness of pipe wall over ³⁄₁₆ inch. (True or False)

47. For forehand welding in the horizontal fixed position, travel begins at the _____ of the pipe.

48. Backhand pipe welding is based on the principle that hot steel will absorb _____, which lowers the melting.

49. When the backhand technique is used, vertical pipe welding can travel from top to bottom. (True or False)

CHAPTER 9

Braze Welding and Advanced Gas Welding Practice: Jobs 9-J39–J49

Please answer the following questions by choosing the letter of the correct answer, circling true or false, or filling in the blanks.

1. All but one of the following is true. Which one is incorrect? Braze welding

    a. is done with a bronze filler rod

    b. does not require fusion of the base metal

    c. requires a special joint design to be used only for torch brazing

    d. joins dissimilar metals

2. All but one of the following is true. Which one is incorrect? Braze welding

    a. is recommended for parts to be used in high temperatures

    b. reduces expansion and contraction

    c. increases welding speed

    d. is the only means of welding malleable cast iron

3. Bronze filler rod melts at _____ degrees F.

4. Allowing a small amount of molten bronze filler rod to spread over a joint is called
   _____.

5. If the base metal produces white smoke during braze welding, the surface is too
   _____.

6. All but one of the following is true. Which one is incorrect? The graphite in gray cast iron

    a. makes machining easy   **b.** lessens tensile strength

    c. limits welding to the fusion method   **d.** causes low ductility

7. Gray cast iron is _____ welded when the color of the base metal must be retained.

8. When fusion welding cast iron, stress caused by expansion and contraction may be reduced by
   _____.

9. Blowholes and porosity in cast iron welds may be caused by the use of too much
   _____.

10. For aluminum, all but one of the following is true. Which one is incorrect?

    a. is lightweight but strong   **b.** is best suited to oxyacetylene welding

    c. conducts electricity well   **d.** resists corrosion

    e. does not change color when heated

11. Aluminum melts at a _____ temperature than copper and steel.

12. Aluminum's tendency to collapse when it reaches its melting point is called _____.

13. Aluminum is one of the metals that has become important to industry. (True or False)

14. The diameter of an aluminum filler rod should equal the _____ metal being welded.

15. Aluminum expands during heating. For this reason, aluminum welds have a tendency to _____ because of the shrinkage that takes place in the weld metal when cooling.

16. All but one of the following is true. Which one is incorrect? Powdered aluminum-welding flux
    a. may be mixed with alcohol or water
    b. should coat the welding rod used on sheet aluminum
    c. is applied to the end of filler rod used for cast aluminum
    d. must be mixed in steel containers to avoid contamination

17. A soft flame is essential in all oxy-gas welding of aluminum. (True or False)

18. All aluminum joints require edge preparation before welding. (True or False)

19. Aluminum welding may be done in all positions, but the _____ position is preferred.

20. Flux residues on gas-welded sections may _____ aluminum.

21. Aluminum welds may be steam cleaned or dipped in a _____ bath.

22. Stirring the filler rod in the aluminum weld pool reduces porosity by bringing _____ to the surface of the pool.

23. The cone of the flame should touch the weld pool to assist in moving it along. (True or False)

24. Weld cracking is a greater problem in welding sheet aluminum than cast aluminum. (True or False)

25. Aluminum oxide has a higher melting point than cast aluminum. (True or False)

26. All but one of the following is true. Which one is incorrect? Magnesium
    a. is lighter than aluminum   b. is usually alloyed with aluminum
    c. is strongest used in its pure state   d. resists corrosion

27. When welding magnesium with the oxyacetylene process, all but one of the following is correct. Which one is incorrect?

    a. avoid lap and fillet welds

    b. support joints to prevent collapse

    c. use flux sparingly

    d. preheat for easier removal of oxide film

    e. treat finished weld with sodium dichromate

28. Three of the following are true. Which one is incorrect? Lead

    a. is malleable and ductile   **b.** is combined with antimony for strength

    c. conducts electricity   **d.** withstands corrosive liquids

29. Problems encountered when oxyacetylene welding lead include three of the following. Which is not true?

    a. difficult edge preparation   **b.** proper torch technique

    c. excess fusion   **d.** burn-through

30. For hard facing a metal surface exposed too much wear, three of the following are true. Which one is incorrect?

    a. adds no new qualities to the base metal

    b. increases service life as much as 40 times

    c. allows rebuilding rather than replacement

    d. allows the use of cheaper materials for the part being surfaced

31. Match the hard-facing welding processes to the results they achieve.

    1. Oxyacetylene   **a.** Flawless surface

    2. Gas tungsten-arc   **b.** Light surface coating on heavy sections

    3. Shielded metal-arc   **c.** Suitable surface for most applications

    4. Atomic hydrogen   **d.** Heavy metal deposit

32. Three of the following hard-surfacing rods are used to produce resistance to abrasion. Which one is not?

    a. cobalt baser   **b.** iron base   **c.** copper base   **d.** tungsten carbide

33. When applying hard-facing materials to steel, all but one of the following are correct. Which one is incorrect?

    a. preheat

    b. use an oxidizing flame to produce sweating

    c. allow the rod to lightly touch the sweating area

    d. use the flame's pressure to move the weld puddle

    e. cool slowly

34. Match the following materials to the special handling they require for hard surfacing.

1. Cast iron
2. Alloy steels
3. High speed steels

a. Use less acetylene in the torch flame

b. Anneal fully and cool very slowly

c. Preheat and post cool carefully to avoid cracking

a. _____

b. _____

c. _____

Name: _____ Date: _____

CHAPTER 10

Soldering and Brazing Principles and Practices: Jobs 10-J50–J51

Please answer the following questions by choosing the letter of the correct answer, circling true or false, or filling in the blanks.

1. Copper pipe and tubing are used for plumbing and _____ installations.

2. Soldering requires higher temperatures than brazing. (True or False)

3. The flow of a liquid drawn into a small space between wet surfaces is called _____ action.

4. Capillary action is not a factor in the distribution of the brazing filler metal during braze welding. (True or False)

5. Selection of the proper material for soldering depends upon all but one of the following. Which one is incorrect?
   a. the metals to be joined   b. the expected service
   c. the desired color of the solder d. the operational temperature of the joint

6. Soldering must meet which one of the three following criteria?
   a. The parts must be joined without melting the base metals.
   b. The filler metal must have a liquidus temperature below 840°F.
   c. The filler metal must wet the base metal surface and be drawn into or held in the joint by capillary action.
   d. All of the above.

7. Match the types of solders with the most appropriate descriptive phrase.
   1. Tin lead   **a.** Most commonly used   **a.** _____
   2. Tin antimony   **b.** Non-toxic for use in food handling equipment   **b.** _____
   3. Tin zinc   **c.** Makes wide-clearance aluminum joints   **c.** _____
   4. Cadmium silver   **d.** Suitable for butt joints in copper tube   **d.** _____
   5. Cadmium zinc   **e.** In the proper ratio, melts and solidifies at the same **e.** _____
                           temperature

8. Soldering fluxes do all but one of the following. Which one is incorrect?
   a. protect the base metal from oxidation
   b. improve filler metal flow
   c. increase wetting action
   d. clean the base metal
   e. float out oxides in the joint

9. Fluxes may be classified in three groups. Which one of the following is incorrect?

    a. inorganic   **b.** organic   **c.** mineral   **d.** rosin

10. Match the types of fluxes to the phrases which best describe them.

1. Highly corrosive	**a.** Very active but highly volatile	**a.** _____
2. Intermediate	**b.** May be localized at a joint to protect the rest of the work	**b.** _____
3. Noncorrosive	**c.** Deposit chemically active residue on completed joint	**c.** _____
4. Paste	**d.** Favored by electrical industry	**d.** _____

11. A clearance of _____ inch is recommended for capillary attraction to function well.

12. The tensile strength of a joint increase as the joint clearance increases. (True or False)

13. The heat required for soldering may be supplied by all but one of the following. Which one is incorrect?

    a. an electric arc   **b.** a torch   **c.** an oven   **d.** induction   **e.** resistance

14. The lowest flame temperature is produced by _____.

    a. propane   **b.** manufactured gas

    c. butane   **d.** natural gas   **e.** acetylene

15. A sooty flame may prevent the solder from flowing smoothly by depositing _____ on the base metal.

16. Pipe to be soldered or brazed is usually mechanically cleaned. (True or False)

17. A tube may be too clean for effective soldering. (True or False)

18. Production operations usually require _____ cleaning.

19. The flux preferred for soldering copper tubing is _____.

    a. highly corrosive   **b.** intermediate

    c. noncorrosive   **d.** paste

20. When soldering copper tubing _____.

    a. prepare and flux joints at least a day before soldering

    b. heat as large an area as possible

    c. apply flame directly to solder until it is melted

    d. immediately pour water over the finished joint to wash away corrosive flux

21. The entire _____ area of a well-soldered joint will be covered with solder.

22. Poorly soldered areas may be caused by all but one of the following. Which one is incorrect?
    a. removing the torch flame from the solder pool   b. poor cleaning
    c. improper fluxing   d. loose fittings

23. Spelter is used for brazing copper tubing. (True or False)

24. Compared to soldering, brazing does three of the following. Which one is incorrect?
    a. uses higher temperatures   b. requires special fluxes
    c. has less of an annealing effect   d. forms a stronger joint

25. The minimum temperature at which brazing will take place is called the _____ temperature.

26. The American Pipe Society lists classifications of brazing filler metal. (True or False)

27. For brazed joints three of the following are true. Which one is incorrect?
    a. is stronger than threaded ones   b. are 90% as strong as the actual fittings
    c. are not loosened by vibration   d. do not leak

28. Brazing filler metals that require the use of a flux include three of the following. Which one is incorrect?
    a. aluminum silicon   b. copper phosphorous
    c. copper   d. nickel

29. For joining parts in electron tube assemblies, _____ filler metals are used.

30. Except for some metals with low melting points, virtually all metals can be brazed with _____ alloys.

31. At _____ degrees F, flux loses its protective qualities.

32. Hard water cannot be effectively used in a flux. (True or False)

33. Brazing satisfactorily joins only similar metals. (True or False)

34. Lap joints are stronger than butt joints. (True or False)

35. Filler metals used for brazing are better electrical conductors than the copper tubing they join. (True or False)

36. Use _____ joints to fabricate pressure-tight assemblies.

37. The lowest flame temperatures are achieved with _____.

    a. air-gas torches   **b.** butane torches

    c. ox-hydrogen torches   **d.** oxyacetylene torches

38. The widest range of heat control is provided by _____ torches.

39. Only brazing filler metals, which may be used with _____, are suitable for torch heating.

40. It is difficult to braze cast iron. (True or False)

41. When fluxing copper tubing in preparation for brazing, do all of the following except _____.

    a. apply flux only in brazing area

    b. apply flux immediately after cleaning

    c. coat the tube inside and out

    d. always apply with a clean brush

42. Copper tubing has reached the brazing temperature when the flux starts to bubble. (True or False)

43. Wrought copper fittings may be cooled quickly after brazing. (True or False)

44. All brazing fluxes must be removed after the brazing alloy has set. (True or False)

45. A pipe that is too large to heat adequately may be divided into _____ for brazing.

46. _____ destroys the effectiveness of the flux.

CHAPTER 11

Shielded Metal Arc Welding Principles

Please answer the following questions by choosing the letter of the correct answer, circling true or false, or filling in the blanks.

1. In shielded metal-arc welding, the molten weld pool is protected by a gaseous shield and a covering of _____.

2. In shielded metal-arc welding, the heat is generated by an electric arc established between a _____ and the work.

3. The nature of the metal electrode alone determines the chemical and physical characteristics of a shielded metal-arc weld. (True or False)

4. The shielded arc process is not suitable for welding metals with low melting points, such as three of the following. Which one is incorrect?

   a. zinc   **b.** lead   **c.** stainless steel   **d.** tin

5. The most dominant markets for SMAW usage are _____ and Maintenance and Repair.

6. The temperature of the electric arc has been measured as high as _____ degrees F.

7. Identify the elements of the welding circuit shown in Fig. 11-1.

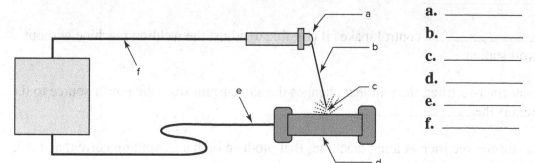

   a. _____
   b. _____
   c. _____
   d. _____
   e. _____
   f. _____

Fig. 11-1.

8. The output slope of welding machines may be classed as either constant voltage or constant _____.

9. Which of the following is not a type of welding machine?

   a. a.c. transformers   **b.** volt-amperage rectifiers

   c. transformer-rectifiers   **d.** engine-driven generators

   e. inverter

10. Inverter welding machines have no disadvantage. (True or False)

11. When the arc is struck and the welding load is on the machine, open circuit voltage drops to _____ voltage at the output terminals of the machine.

12. Lengthening the arc _____
   a. increases both arc voltage and current
   b. decreases both arc voltage and current
   c. increases arc voltage and decreases current
   d. decreases arc voltage and increases current

13. Engine generator welding machines can produce both types of welding current. (True or False)

14. Machine sizes are determined on the basis of _____.

15. The actual output of a welding machine is always higher than the rated output. (True or False)

16. A _____ switch is used to change the direction of the flow of welding d.c. current.

17. Volt-ampere meters may indicate polarity. (True or False)

18. Voltage measures the _____ of an electric current.

19. Amperage measures the _____ of an electric current.

20. The ability to strike and maintain a constant arc is determined by a welding machine's
_____.

21. _____ control makes it possible to adjust the welding machine without leaving the workstation.

22. On a d.c. transformer-rectifier, the unit that changes the a.c. current from the power source to d.c. welding current is the _____.

23. There are transformer-rectifier welding machines that produce both a.c. welding current and d.c. welding current. (True or False)

24. The a.c. welding machines have gained in popularity because of advantages such as those listed below. Which one is incorrect?
   a. improved machine performance
   b. noiseless operation
   c. adaptability to the use of large electrodes
   d. decreased power consumption
   e. adaptability to both straight and reverse polarity

25. The _____ welding machine will only produce a.c. current.

26. The most important consideration in determining which type of welding machine to buy is the initial purchase cost. (True or False)

27. To complete the circuit between a welding machine and the work, an _____ cable and a work cable are required.

28. Welding cable should be stiff so that it will wear well. (True or False)

29. To determine the size of cable to choose for welding, consider machine _____ and distance from the work.

30. Resistance to the flow of the welding current increases as the diameter of the cable increases. (True or False)

31. The electrode holder used for shielded metal-arc welding may also be used for carbon-arc welding. (True or False)

32. Whenever in the working area, it is necessary to wear _____ as well as under the welding hood.

33. Safety glasses can absorb more than 99.9 percent of harmful ultraviolet rays. (True or False)

34. Chrome _____ helmets are used in cramped areas where there is not enough room for a standard hard helmet.

35. Electronic filter welding helmets are not useful when working in close quarters. (True or False)

36. Match the pieces of welding equipment to their best descriptions.

1. Electrode holder	**a.** Connects cable to welding machine	**a.** _____
2. Cable	**b.** Supplies the weld metal	**b.** _____
3. Coating and/or Electrode core	**c.** Can produce direct or alternating current	**c.** _____
4. Rotary clamp	**d.** Turns with work without twisting cable	**d.** _____
5. Lug	**e.** Changes alternating current to direct current	**e.** _____
6. Rectifier	**f.** Carries electric current from the welding cable to the electrode	**f.** _____
7. Transformer-rectifier machine	**g.** Completes electrical circuit between the welding machine and the work	**g.** _____

37. Match the welding terms to their best descriptions.

1. Penetration	**a.** Volt-ampere characteristic	**a.** _____
2. Voltage	**b.** Generates heat for welding	**b.** _____
3. Amperage	**c.** Depth of the weld into the plate	**c.** _____
4. Arc	**d.** Measure of the amount of electrical current	**d.** _____
5. Output slope	**e.** Completes electrical circuit to protect welder from accidental shock	**e.** _____
6. Ground	**f.** Force with which electrical current flows	**f.** _____

38. The clothing of choice for welders should be 100 percent cotton or wool. (True or False)

CHAPTER 12

Shielded Metal Arc Welding Electrodes

Please answer the following questions by choosing the letter of the correct answer, circling true or false, or filling in the blanks.

1. Shielded metal-arc welding may be referred to as all but one of the following. Which is incorrect?

   a. arc welding   **b.** SMAW   **c.** laser welding   **d.** stick electrode welding

2. _____ is composite filler metal electrode consisting of a core of a bare electrode or metal cored electrode to which a covering sufficient to provide a slag layer on the weld metal has been applied.

3. The type of electrode covering influences the degree of penetration and the depth of the crater. (True or False)

4. The addition of large amounts of iron power to the coating of an electrode increases the speed of welding and improves the weld appearance. (True or False)

5. As a rule, when welding with a SMAW electrode, the maximum arc length is never greater than the diameter of the bare end of the electrode. (True or False)

6. Molten slag not only protects the weld metal, but also does three of the following. Which is incorrect?

   a. removes oxides and impurities

   b. speeds up the cooling rate of the molten metal

   c. slows down the cooling rate of the solidified weld metal

   d. controls the shape and appearance of the weld

7. The electrode covering may be used to introduce additional elements into the weld metal. (True or False)

8. The welding of high carbon and alloy steels, high sulfur steels, and phosphorous-bearing steels has been improved by the introduction of low _____ electrodes.

9. The covering on a welding electrode makes it possible for four of the following. Which one is incorrect?

   a. to start the arc more easily   **b.** to vary arc length

   c. to concentrate the heat of the arc on the electrode   **d.** work with higher currents

   e. to control the arc more effectively

10. To a large extent the core-wire composition of mild steel electrodes is changed depending upon the kind of covering used on the electrode. (True or False)

11. The following electrode coverings are correctly paired with the purposes they serve. Which one is incorrect?

   a. ferro-alloys: deoxidizers

   b. alkaline earth metals: provide reducing gases

   c. silica, clay: fluxes

   d. iron ore, mica: slagging ingredients

   e. potassium silicates: binders

12. The composition of the covering determines the best polarity to use for d.c. applications. (True or False)

13. The first two numbers in the AWS electrode classification indicates _____.

   a. polarity   **b.** tensile strength

   c. welding current   **d.** position to welding

   e. covering

14. The next to the last number in the AWS electrode classification indicates

   _____

   a. polarity   **b.** covering

   c. position of welding   **d.** tensile strength

   e. welding current

15. The last number in the AWS electrode classification indicates the type of

   _____ to be used and the covering on the electrode.

16. Selecting the proper size of electrode to use on a job depends upon all but which of the following?

   a. joint design   **b.** material thickness

   c. size of welding machine   **d.** thickness of weld layers

   e. welding position required

17. The covering should have a melting point lower than that of both the core wire and the base metal when welding in the flat position. (True or False)

18. Certification in the F4 group designation does not qualify in any other group. (True or False)

19. Match the electrode operating characteristics to their best descriptions.

   1. Fast fill   **a.** Deposits weld metal which solidifies rapidly   **a.** _____

   2. Fast follow   **b.** Designed for fast downhand welding and easy   **b.** _____
                               slag removal

   3. Fast freeze   **c.** Produces ductile welds free from underbead and   **c.** _____
                               microcracking

   4. Low hydrogen   **d.** Contributes additional metal to weld deposit   **d.** _____

   5. Iron powder   **e.** Designed for vertical position, travel-down   **e.** _____
                               welding and reduced burn-through

20. Match the types of electrodes to their operating characteristics.

1. EXX13	**a.** Fast fill	**a.**	_____
2. EXX24	**b.** Low hydrogen	**b.**	_____
3. EXX10	**c.** Fill freeze	**c.**	_____
4. EXX18	**d.** Fast freeze	**d.**	_____

21. Of the following metals, the one with the lowest melting point is _____.

   a. zinc   **b.** steel   **c.** copper   **d.** silver   **e.** lead

22. According to textbook Tables 12-10 through 12-12, for cast iron three of the following are true. Which one is incorrect?

   a. forms tough but breakable slag

   b. melts at a moderate speed

   c. can be made to form continuous chips

   d. makes a molten puddle that does not spark under blowpipe flame

23. In general never use an electrode having a diameter _____ than the thickness of the material being welded.

24. For production work, the largest electrode size that can be handled should be used. (True or False)

25. For overhead, vertical, or horizontal welding, EXX14 or EXX24 electrodes may be used. (True or False)

26. For easier welding in the horizontal and flat positions, use electrodes classified in all but one of the following.

   a. EXX13   **b.** EXX14   **c.** EXX27   **d.** EXX28   **e.** EXX29

27. To increase welding speed, use electrodes with _____ diameters.

28. Match the electrode classifications to their best descriptions.

1. E6010	**a.** Covering contains easily ionized materials	**a.**	_____
2. E6011	**b.** Best adapted for vertical and overhead welding	**b.**	_____
3. E7014	**c.** Suitable for welding mild steel in all positions	**c.**	_____
4. E6013	**d.** Requires alternating current to retain its best characteristics	**d.**	_____

29. Tensile strength exceeding 100,000 psi may be obtained by welding with a covered electrode with an alloy steel core wire. (True or False)

30. Stainless steel electrodes may contain both chromium and _____.

31. For stainless steel electrodes, four of the following are true. Which is incorrect?

   a. resist corrosion

   b. withstand high temperatures

   c. do not oxidize readily

   d. are easily cold worked

   e. resist scaling

32. Match the stainless steel classifications to their best descriptions.

   1. Martensitic   **a.** Requires no heat treatment during welding   **a.** _____

   2. Ferritic   **b.** Magnetic, normally soft and ductile   **b.** _____

   3. Austenitic   **c.** Air hardening, normally hard and brittle   **c.** _____

33. SMAW electrodes are available for all PH stainless steel types. (True or False)

34. _____ is the deposition of an alloy material on a metal part to form a protective surface.

35. Copper-nickel (ECuNi) electrodes are used with _____ polarity.

36. E1100 and E4043 electrodes are the most commonly used covered electrodes for welding aluminum. (True or False)

37. Hot-shortness is a problem in the welding of aluminum. (True or False)

38. The presence of _____ in the electrode covering is a major cause of a porous weld, underbead cracking, and rough appearance.

39. All mineral-covered electrodes are thirsty. The minute they are unpacked, they start absorbing moisture—too much moisture for a sound weld. (True or False)

40. Although it is recommended, oven storage is not essential for preserving the quality of hard-surfacing electrodes. (True or False)

CHAPTER 13

Shielded Metal Arc Welding Practice: Jobs 13-J1–J25 (Plate)

Please answer the following questions by choosing the letter of the correct answer, circling true or false, or filling in the blanks.

1. SMAW accounts for less than 50% of the total amount of welding being done. (True or False)

2. For stick electrode welding, all but one of the following are true. Which one is incorrect?
   a. versatile
   b. dependable
   c. expensive but fast
   d. applicable to all welding positions

3. When welding, be as comfortable as possible in order to avoid _____.

4. Electrode positive will produce the deepest penetrating weld. (True or False)

5. In a d.c. circuit, the hottest side is the one with positive polarity. (True or False)

6. When welding is performed in electrode negative, the electrode lead becomes the negative side of the circuit and the work lead becomes the positive. (True or False)

7. A fairly short arc is preferred for metal-arc welding in all but one of the following. Which one is incorrect?
   a. concentrates heat on the work
   b. reduces arc blow
   c. protects the weld pool from contamination
   d. blows out the flame periodically to rest the welder

8. In comparison to other welding positions, a slightly longer arc may be maintained for welding that is _____.
   a. vertical   **b.** flat   **c.** horizontal   **d.** overhead

9. As arc length increases, arc voltage _____.

10. The type of coating on the electrode may affect the arc length. (True or False)

11. The arc is too long in all but one of the following. Which one is incorrect?
    a. the metal flutters
    b. slight explosions occur regularly
    c. there is a sharp, crackling sound
    d. the metal is easily seen

12. For arc blow, all but one of the following are true. Which one is incorrect?

    a. is caused by magnetic forces   **b.** is more of a problem with a short arc length

    c. may cause porosity   **d.** may be controlled by changing the electrode's position

13. The _____ equivalent of the fractional electrode sizes gives some idea of the heat setting.

14. A ⅛-inch electrode requires higher current than a ⁵⁄₃₂-inch electrode. (True or False)

15. Use more current for all but one of the following. Which one is incorrect?

    a. good heat conductors   **b.** electrode negative electrodes

    c. vertical-position welding   **d.** larger pieces of material

16. Current requirements vary with welding speed. (True or False)

17. Holding the electrode at a proper _____ to the work prevents undercut and slag inclusions.

18. Practice with covered electrodes allows a student to do all but which one of the following?

    a. observe what takes place in the arc

    b. avoid distraction caused by arc blow

    c. observe fusion of weld metal and bare metal

    d. develop a steady hand for maintaining arc gap

19. Match the following welding problems to possible methods of correcting them.

    1. Spatter   **a.** Choose an electrode small enough to reach the   **a.** _____
bottom of narrow V's

    2. Undercut   **b.** Paint parts next to weld area   **b.** _____

    3. Poor fusion   **c.** Use a.c. current if possible   **c.** _____

    4. Porosity   **d.** Make a pool to keep the weld molten longer   **d.** _____

    5. Arc blow   **e.** Avoid creating too large a weld pool with an   **e.** _____
oversized electrode

20. To maintain the flow of electric current, the electrode should be kept lightly touching the plate. (True or False)

21. Traveling too fast when making continuous stringer beads produces beads with all but one of the following. Which one is incorrect?

    a. are narrow   **b.** show little reinforcement

    c. overlap   **d.** have coarse ripples

22. One technique to fill a crater is to _____ the arc a short time over the crater while maintaining the proper arc length.

23. For deep groove joints, deposit beads wider than stringer beads by _____ the electrode from side to side.

24. Use a _____ weld to make a lap joint.

25. Large fillets are often welded with multipass stringer beads. (True or False)

26. The strength of a fillet weld depends upon its _____ thickness.

27. Welds made with covered electrodes are as strong and ductile as the base metal. (True or False)

28. Unprotected molten metal combines with gases in the air to form oxides and _____.

29. Shielded-arc electrodes produce weld metal that is superior to the base metal. (True or False)

30. Arc blow is a problem with covered electrodes because of the covering on the electrode. (True or False)

31. Industry tends to use larger electrodes than it is recommended to begin practicing with. (True or False)

32. E6013 electrodes are not suited to weld thin metals. (True or False)

33. It is easier to weld in the vertical and overhead positions with a.c. electrodes. (True or False)

34. Compared to direct welding current, all but one of the following are true of alternating current. Which one is incorrect?
    a. avoids arc blow
    b. uses larger electrodes
    c. strikes an arc more easily
    d. welds heavy steels faster

35. Too short an arc gap causes _____.
    a. rough welds
    b. slag inclusions
    c. spatter
    d. oxidation

36. The crater can be filled only with one technique. (True or False)

37. The edge joint is an economical joint to set up. (True or False)

38. It is not good practice to weave beads wider than _____ times the diameter of the coated electrode.

39. For E6010 electrodes, all but one of the following are true. Which one is not true?
    a. produce high quality welds in all positions
    b. have a deep penetrating, spray-type arc
    c. produce no spatter
    d. form a quick-setting weld pool

40. E6011 electrodes produce weld metal with higher ductility and tensile strength than that produced by E6010 electrodes. (True or False)

41. The arc gap for the E6010 electrode (electrode positive) may be less than for the E6013 (electrode negative) electrode. (True or False)

42. Industrial weaved beading is usually done with _____ electrodes.

43. Compared to DCEN, for E6013 electrodes three of the following are true. Which one is incorrect?
    a. require higher bead reinforcement   b. penetrate deeper
    c. form less slag   d. solidify more quickly

44. Edge joints are used extensively for high-pressure vessel work. (True or False)

45. For edge joints, three of the following are true. Which one is incorrect?
    a. strong   b. economical   c. easy to weld   d. a.c. or d.c. electrodes can be used

46. The edge joint will not stand very much load when subjected to tension and bending. (True or False)

47. For welding an edge joint, an electrode should be held at a 90° angle to the plate. (True or False)

48. The lap joint is strongest when a double fillet weld is used. (True or False)

49. Lap joints require little preparation and fitup. (True or False)

50. The top edge of the top plate of a lap joint has a tendency to burn away. (True or False)

51. Travel that is too _____ will deposit excess weld metal on the plate.

52. E6010 electrodes have a greater tendency to undercut than E6013 electrodes. (True or False)

53. Lap joints are welded with _____ welds in various positions.

54. Stringer beads are used to weld horizontal _____ welds and fillet welds.

55. For making horizontal stringer beads, the plate is in the _____ position.

56. In contrast to flat welding of stringer beads, horizontal beading requires all but which of the following?

    a. lower current   **b.** longer arc

    c. electrode movement to avoid sagging   **d.** good fusion between beads

57. Welding downhill on a vertical surface is used for critical pipeline and sheet-metal work. (True or False)

58. Vertical down welding requires a _____ arc gap.

59. Downhill vertical welding requires a fast rate of travel. (True or False)

60. Downhill vertical lap joint welds are used when maximum reinforcement and strength are required. (True or False)

61. Undercutting is not a problem when downhill vertical welds are made with DCEN. (True or False)

62. For tank and structural work, it is often necessary to make fillet welds in the _____ position.

63. A T-joint is typically welded with a _____ weld.

64. For multipass fillet welds, no more than _____-inch thickness of weld metal should be deposited on each layer.

65. Most multipass welding is done with DCEP electrodes. (True or False)

66. When making a weaved fillet weld with E6010 electrodes, hesitate longer at the sides of the weld than you would with electrode negative electrodes. (T or F)

67. Refer to the print in Fig. 13-1 on the next page to match the measurements to the appropriate descriptions.

1. Plate thickness	**a.** $1\frac{1}{2}''$	**a.** _____
2. Plate # 7 width	**b.** $\frac{1}{4}''$	**b.** _____
3. Plate # 6 width	**c.** $\frac{1}{2}''$	**c.** _____
4. Leg length of second fillet pass	**d.** $\frac{3}{8}''$	**d.** _____
5. Leg length of third fillet pass	**e.** $\frac{3}{16}''$	**e.** _____
6. Size of electrode used	**f.** $3\frac{1}{4}''$	**f.** _____

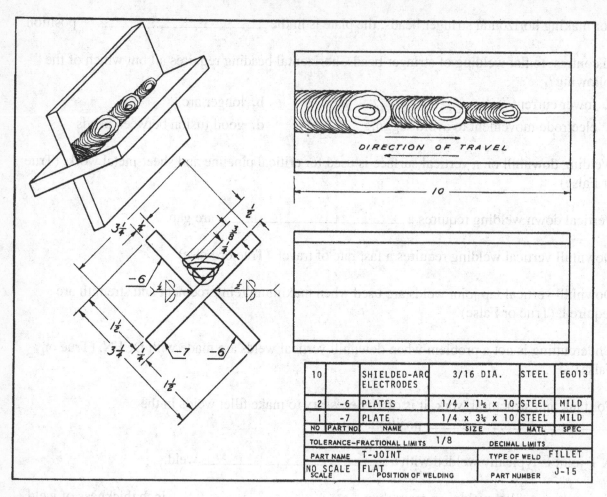

The table in the figure:

NO	PART NO	NAME	SIZE	MATL	SPEC
10		SHIELDED-ARC ELECTRODES	3/16 DIA.	STEEL	E6013
2	-6	PLATES	1/4 x 1½ x 10	STEEL	MILD
1	-7	PLATE	1/4 x 3½ x 10	STEEL	MILD

TOLERANCE–FRACTIONAL LIMITS 1/8		DECIMAL LIMITS	
PART NAME T-JOINT		TYPE OF WELD FILLET	
SCALE NO SCALE	FLAT POSITION OF WELDING	PART NUMBER	J-15

Fig. 13-1.

68. Interrupt multipass fillet welding to remove slag between each pass. (True or False)

69. Electrode negative electrodes are generally used for vertical welding with travel up. (True or False)

70. When making a vertical weld with E6010 type electrodes, lengthen the arc gap on the upward stroke to allow deposited weld metal to solidify. (True or False)

71. When welding in the vertical position calls for weaving a bead, travel is usually down. (True or False)

72. If electrode negative electrodes are used for weave beading in the vertical position with upward travel, speed of travel should be _____ than with reverse polarity.

73. Much of the welding done in the field is _____ fillet welding.

74. The electrodes recommended for the job in Fig. 13-2 are E _____ or E _____.

75. The electrode sizes recommended for the job in Fig. 13-2 include all but which of the following.

    a. ¼″   **b.** ⅛″   **c.** ³⁄₁₆″   **d.** ⁵⁄₃₂″

52   **STUDENT WORKBOOK**

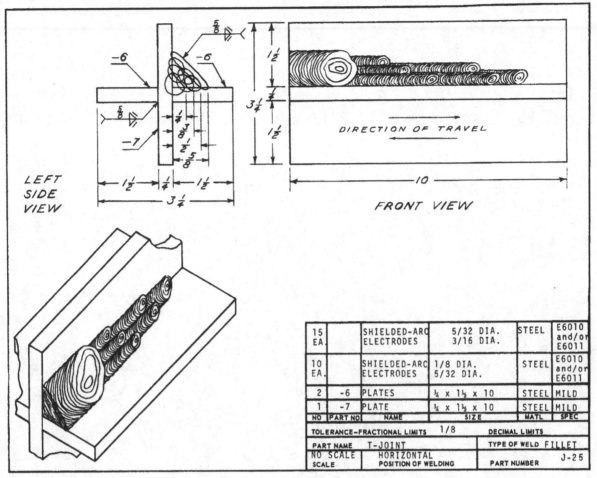

Fig. 13-2.

76. Refer to Fig. 13-2 to match the measurements to the appropriate descriptions.

1. Limits of tolerance in fractions	a. ¼″	a. _____
2. Print number	b. ⅛″	b. _____
3. Size of first fillet weld	c. ⅜″	c. _____
4. Size of passes 2 and 3	d. ⅝″	d. _____
5. Size of passes 4, 5, and 6	e. ½″	e. _____
6. Size of lacing bead	f. J-25	f. _____

77. The print in Fig. 13-2 calls for a _____ welding position.

78. The part to be welded in Fig. 13-2 is a _____.

Fig. 13-2.

76. Refer to Fig. 13-2 to match the measurements to the appropriate descriptions.

1. Limits of tolerance in fractions _____ a. _____

2. Print number _____ b. _____

3. Size of first filler weld _____ c. _____

4. Size of passes 2 and 3 _____ d. _____

5. Size of passes 4, 5, and 6 _____ e. _____

6. Size of facing bead _____ f. _____

77. The print in Fig. 13-2 calls for a _____ welding position.

78. This part to be welded in Fig. 13-2 is a _____

CHAPTER 14

Shielded Metal Arc Welding Practice: Jobs 14-J26–J42 (Plate)

Please answer the following questions by choosing the letter of the correct answer, circling true or false, or filling in the blanks.

1. For economic reasons and quality reasons, in the shop every effort is made to position the work so that welding can be in the flat or horizontal positions. (True or False)

2. A greater amount of out-of-position welding is encountered in field work. (True or False)

3. If you are standing, drape the heavy welding cable over your shoulders to reduce the weight on the arm you are welding with. (True or False)

4. The whipping technique should be used with low hydrogen type electrodes (E7016-E7018). (True or False)

5. Backing material to a single-V butt joint creates the effect of welding a joint with a
   _____ ring.

6. Multipass welding of a single-V butt joint with a backing bar requires complete fusion with all but which one of the following?
   a. backing bar
   b. bevel plates
   c. preceding passes
   d. slag remnants

7. Refer to the _____ of the beveled plate to determine the width of the last pass.

8. The completed weld of a single-V butt joint should have a reinforcement of about
   _____ inch.

9. The strength of a square groove butt joint welded from only one side depends upon but which of the following?
   a. penetration
   b. size of electrode used
   c. speed of welding
   d. amount of current
   e. the length of the plate

10. For outside corner joints, which one of the following is not true?

    a. are seldom used because of difficult preparation

    b. require welding conditions similar to those for V-groove butt joints

    c. require full penetration through the backside

    d. are more efficient with an inside corner fillet weld

11. Look at Fig.14-1 to match the following measurements to the appropriate descriptions.

    1. Surface plate thickness   **a.** ³⁄₁₆   **a.** _____

    2. Backup plate thickness   **b.** 10   **b.** _____

    3. Joint length   **c.** 1 ½   **c.** _____

    4. Composite joints' width   **d.** ³⁄₈   **d.** _____

    5. Root, opening   **e.** 7   **e.** _____

    6. Number of stringer beads required   **f.** 8¼   **f.** _____

    7. Width of surface plates   **g.** ⁵⁄₃₂   **g.** _____

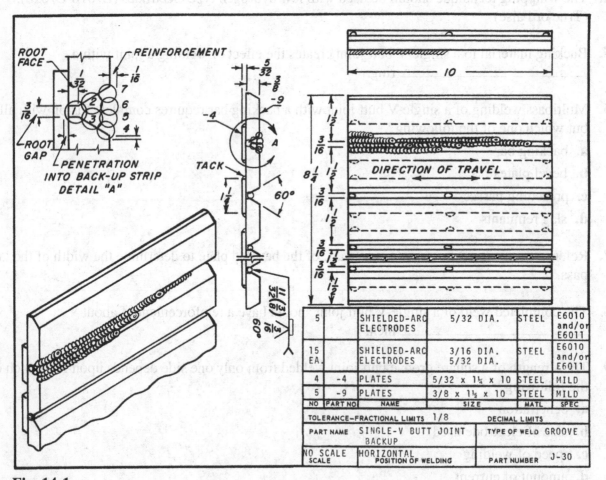

Fig. 14-1.

Name: _____ Date: _____

12. In welding the first pass in a flat position, single-V butt joint all but one of the following are correct. Which one is incorrect?

  a. hesitate at the sides of the weld

  b. keep the electrode within the space provided by the root gap

  c. deposit droplets of metal ahead of the weld

  d. maintain a small hole at the leading edge of the weld crater

  e. form a small bead on the reverse side of the groove

13. The second pass of a single-V butt joint must achieve fusion with both the root pass and the _____ of the plate.

14. All low hydrogen electrodes produce welds that are practically free of _____.

15. For E7016 electrodes, all but one of the following are true. Which one is incorret?

  a. reduce underbead on low alloy steels

  b. are recommended for welds to be porcelain enameled

  c. produce welds suitable for nearly all code work

  d. use preheating to reduce stress requirements

16. For F7016 electrodes, all but one of the following are true. Which one is incorrect?

  a. use DCEP for their larger diameters

  b. are not suitable for vertical welding in their larger sizes

  c. produce rapidly freezing weld metal

  d. have a quiet arc with medium penetration and little spatter

17. Match the measurements from Fig. 14-2 on the next page to the appropriate descriptions.

1. First-pass electrode sizes	**a.** $3/32''-1/8''$	**a.** _____
2. Plate gap	**b.** $1/16''$	**b.** _____
3. Back-side reinforcement on first pass	**c.** $1/8''$ or $5/32''$	**c.** _____
4. Finish-pass reinforcement	**d.** $1/8''$	**d.** _____
5. Finish-pass beyond shoulder	**e.** $3/32''$	**e.** _____

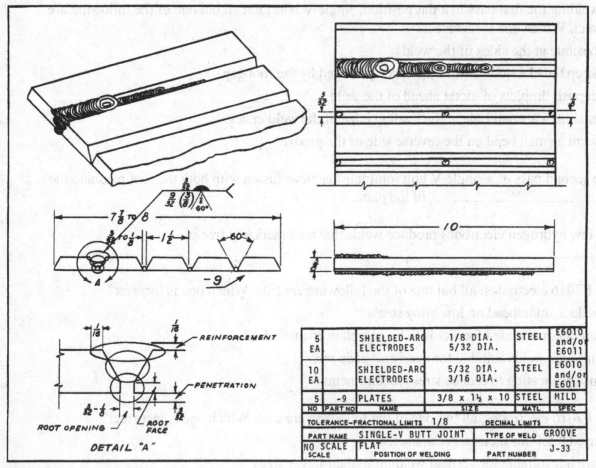

Fig. 14-2.

18. When making a vertical weld, travel up, _____ will make the weld pool spill off and run down.

19. The whipping technique is used to help control the _____ penetration characteristics of cellulose E6010-E6011 electrodes.

20. Multipass, vertical fillet welds are usually welded travel up with _____ polarity.

21. The iron powder in E7018 electrodes gives them the following advantages over the E7016 electrodes except for which of the following?

    a. higher deposition rate

    b. better restrike characteristics

    c. rougher bead

    d. more stable arc

22. Coatings containing iron powder are _____ than regular low hydrogen coatings.

23. Using an electrode with the coating in contact with the work is called _____ technique.

24. Downhill welding is _____ than the uphill method.

25. When downhill welding, keep the weld pool and slag from running ahead of the arc by using a shorter arc and _____ the electrode travel angle and rate of travel.

CHAPTER 15

Shielded Metal Arc Welding Practice: Jobs 15-J43–J55 (Plate)

Please answer the following questions by choosing the letter of the correct answer, circling true or false, or filling in the blanks.

1. It is necessary to read job drawings and sketches to satisfy the skill demands of industrial shielded metal-arc welding jobs. (True or False)

2. Maximum strength in a single-V butt joint can be obtained only if there is complete _____ through the backside of the joint.

3. The whipping motion technique should never be used on low-hydrogen electrodes. (True or False)

4. When making a weaved bead in the vertical position, travel up, avoid excessive convexity in the center of the weld by moving rapidly across the _____ of the weld.

5. When depositing weld metal for a fillet weld in the overhead position, keep a short arc gap. (True or False)

6. A sound fillet weld in a T-joint should break evenly through the _____ when subjected to destructive testing.

7. A multipass fillet weld in the overhead position is usually done with weaved beading. (True or False)

8. Each bead of a multipass fillet must be fused with the preceding _____ as well as the plate surfaces.

9. Each pass of a multipass weld requires a change in the _____ of the electrode.

10. A lap joint in the overhead position must be welded with stringer beads. (True or False)

11. E7018 lap joints in the overhead position may be welded with DCEP or DCEN electrodes. (True or False)

12. Although downhill welding of single-V butt joints is a rather difficult technique, once mastered it is _____ than other methods.

13. V-Groove butt joints in pipe can be welded faster if a _____ is used.

14. For the first pass of a single-V butt joint, travel down, all but one of the following are true. Which one is incorrect?

    a. requires a close arc   **b.** may be worked with a drag technique

    c. must be weaved   **d.** should produce no holes or burn-through

15. When welding a multipass single-V butt joint, travel down, which one of the following is correct?

    a. undercutting may be a problem

    b. hesitate at the sides of the weld to avoid undercut

    c. use electrode negative

    d. make the final bead somewhat convex

16. Holes or voids are acceptable in the second pass of a multipass single-V butt joint, travel down. (True or False)

17. When welding a T-joint in the overhead position with multipass fillets, break the arc and go back to the crater if it begins to solidify. (True or False)

18. A beveled-butt joint welded in the horizontal position is usually made with

_____.

    a. laced beads   **b.** stringer beads

    c. weave beads   **d.** alternating stringer and weaved beads

19. Complete penetration is secured by using the correct combination of all but which one of the following?

    a. current   **b.** arc gap   **c.** electrode position   **d.** travel speed

20. Welding a coupling to flat plate requires an electrode angle that is constantly

_____ in relation to the welder.

21. When welding with a d.c. current, arc blow is not a possibility. (True or False)

22. Match the measurements from Fig. 15-1 to the appropriate descriptions.

    1. Smaller pipe diameter   **a.** ¼″   **a.** _____

    2. Larger pipe diameter   **b.** 5½″   **b.** _____

    3. Size of fillet on smaller pipe   **c.** 10″   **c.** _____

    4. Size of fillet on larger pipe   **d.** 1¼″   **d.** _____

    5. Distance from larger pipe center to nearest plate edge   **e.** 5″   **e.** _____

    6. Center-to-center distance of the larger couplings   **f.** 2¼″   **f.** _____

    7. Length and width of flat plate   **g.** ⅛″   **g.** _____

    8. Fractional tolerance limits   **h.** ³⁄₁₆″   **h.** _____

    9. Distance from the center of the small, centered pipe to   **i.** 3″   **i.** _____
        the plate edges

Name: _____ Date: _____

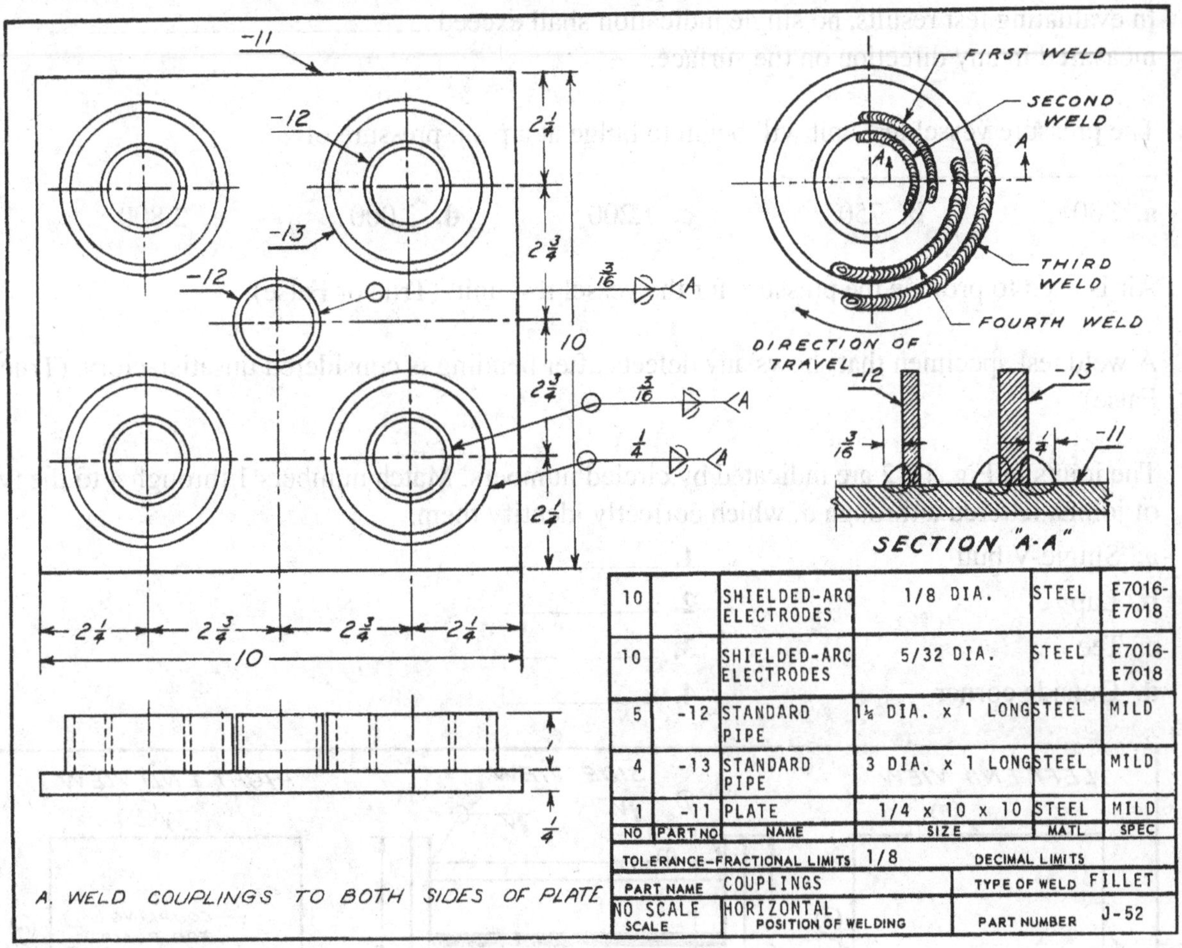

Fig. 15-1.

23. A beveled-butt joint in plate with a backup strip is often set up with a root gap of

_____.

   a. ¾″   **b.** ⁵⁄₁₆″   **c.** ¼″   **d.** ⅜″

24. A single-V butt joint in the overhead position is usually welded with a electrode negative electrodes. (True or False)

25. Weld test specimens are usually made of _____ steel.

26. According to AWS, if a welder qualifies in one electrode group, he or she is also qualified in any lower group F-number. (True or False)

27. A single-V butt joint is used to test _____ welds.

28. In order to qualify, a welder must pass the tests in all positions. (True or False)

29. Groove welds are subjected to face and root _____ tests.

30. Test specimens are bent in a _____ to determine the soundness of the welds.

31. In evaluating test results, no single indication shall exceed _____, measured in any direction on the surface.

32. The pressure vessel test unit will begin to bulge at a p.s.i. pressure of _____.

   a. 300   **b.** 750   **c.** 1,200   **d.** 2,000   **e.** 2,800

33. Air is used to provide the pressure for the vessel test unit. (True or False)

34. A weld test specimen that shows any defects after bending is considered unsatisfactory. (True or False)

35. The joints in Fig. 15-2 are indicated by circled numbers. Match numbers 1 through 4 to the types of joints, lettered a through d, which correctly identify them.

   a. Single-V butt   **1.** _____

   b. Lap   **2.** _____

   c. Tee   **3.** _____

   d. Outside corner   **4.** _____

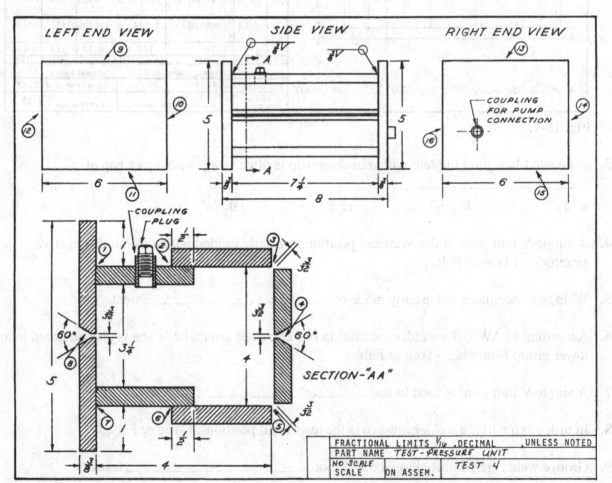

Fig. 15-2.

36. Match the following measurements from Fig. 15-2 to the appropriate dimensions.

1. Length of end plates   **a.** 8″   **a.** _____
2. Width of end plates   **b.** 7¼″   **b.** _____
3. Distance between end plates   **c.** 6″   **c.** _____
4. Overall unit length   **d.** 5″   **d.** _____

CHAPTER 16

Pipe Welding and Shielded Metal Arc Welding Practice: Jobs 16-J1–J17 (Pipe)

Please answer the following questions by choosing the letter of the correct answer, circling true or false, or filling in the blanks.

1. Pipe diameters over 20 inches cannot be welded satisfactorily. (True or False)

2. All but which one of the following ferrous metals are used to make pipe?
   a. nickel **b.** chrome-moly **c.** stainless steel **d.** carbon steel

3. Nonferrous metals used in pipe construction include all but which one of the following?
   a. titanium **b.** copper **c.** asbestos **d.** aluminum

4. Power plants rely heavily on piping systems, pressure vessels, and various other welded structures. (True or False)

5. The most common use for welded piping in buildings is hot water systems for
   _____.

6. Pipelines must be able to service low and high pressures and _____.

7. There are more miles of pipeline in the U.S. than there are miles of railroad track. (True or False)

8. At least _____ % of the nation's overland cargo flows through pipelines.

9. Welded pipeline forms an underground network for transporting only petroleum products. (True or False)

10. Product pipelines transport more than _____ refined petroleum products.

11. It is possible to move a product through a pipeline with an average speed of three to eight miles per hour. (True or False)

12. Sections of pipe may be as long as _____ feet.

13. Round rubber balls or similar separators used to keep different products flowing through pipelines apart are called _____.

14. At the present time, practically all pipe 2 inches in diameter and larger are joined by welding. (True or False)

15. All but one of the following are advantages of welded pipe installations. Which one is incorrect?
    a. permanence **b.** low maintenance costs
    c. heavier, more durable fittings **d.** improved flow characteristics

16. If all the joints on a piping system are welded, the cost of materials alone may be reduced by _____%.

17. Welded pipe fittings allow the use of _____ pipe walls with no loss of efficiency.

18. The life expectancy of a welded pipe system is _____ times that of a threaded system.

19. Thread cutting reduces pipe strength. (True or False)

20. Inside diameters of welded fittings and pipe meet exactly to improve flow. (True or False)

21. Compared to threaded systems, welded piping includes all but which one of the following?
   a. may be shop-fabricated
   b. requires slightly more equipment
   c. requires less space
   d. is more efficient

22. The material most widely used in piping systems is _____.
   a. aluminum
   b. carbon steel
   c. polyvinyl chloride
   d. wrought iron

23. High pressure transmission lines are made of low _____ steel pipe.

24. Chrome-moly pipe is preferred for all but which of the following?
   a. high temperature strength
   b. inhibits graphitization
   c. non air hardening
   d. inhibits oxidization

25. For extremely low or high temperatures, use _____ pipe.

26. A special type of arc welding machine is required to weld pipe. (True or False)

27. For general pipe welding, _____ electrodes are most commonly used.

28. A type of electrode especially designed for vertical down welding belongs to the classification _____.
   a. E6010
   b. E6011
   c. E7010
   d. E11018G

29. The highest tensile strength is achieved in welds made with electrodes belonging to the classification _____.
   a. E6010
   b. E6011
   c. E7010
   d. E11018G

30. Pipe welds may achieve strengths as high as _____ p.s.i.

31. Which one of the following is not one of the joints most commonly used in welded piping systems.
   a. circumferential butt
   b. lap
   c. socket
   d. lateral

32. The types of welds most commonly used in welded piping systems are grooves and
_____.

33. Inner liners used as supports for butt joints may be called chill or _____
rings.

34. All but which of the following are true of pipe butt joints?
    a. are easily prepared   **b.** may be welded in all positions
    c. are faultless except for some cracking   **d.** allow good stress distribution

35. Groove welds are usually applied to pipe for all but which of the following?
    a. sockets   **b.** flanges   **c.** valves   **d.** branches

36. Use of the butt joint for pipe is limited to thick pipe. (True or False)

37. The accepted bevels for welded industrial pipe range from 30 to _____
degrees.

38. For socket joints, all but one of the following are true. Which one is incorrect?
    a. are fillet welded   **b.** are restricted to use in small pipe
    c. are recommended for radioactive systems
    d. are not suitable for exposure to corrosive materials

39. For factory-manufactured welding pipe fittings, all but one of the following are true. Which one is incorrect?
    a. equal the strength of the pipe being welded   **b.** are welded in the horizontal position
    c. require only butt joints (groove welds)   **d.** require less space than fabricated ones

40. Chill rings assist the welder in securing complete penetration and fusion without
_____.

41. Backing rings prevent the formation of icicles or _____ inside the joint.

42. Backing rings that expand or contract to fit the inside diameter of a pipe are called
_____.

43. Backing rings may have a series of small nubs around the ring to aid in spacing the pipe joint.
(True or False)

44. A consumable _____ ring improves weld quality, resists shrinkage and
hot shortness, and eliminates weld-root notches.

45. Fabricated weld fittings are formed from straight lengths of pipe by hand or machine
_____ cutting.

46. Piping codes have been established to insure uniform practices in design, installation, and
_____.

47. Some insurance companies, manufacturers, and the _____ set up their own welding fabrication codes.

48. Match the welding societies to the types of welding for which they provide codes.

1. American Society of Mechanical Engineers   a. Arc welding of piping for transmission, compressing and pumping fuel gases   a. _____

2. American Petroleum Institute   b. Pioneer of training and testing practices   b. _____

3. American Water Works Association   c. Piping connected to boilers   c. _____

4. Heating, Piping, Air Conditioning Contractors   d. Piping for water purification plants   d. _____

5. American Welding Society   e. Piping for heating   e. _____

49. Procedure qualification tests establish a procedure for producing welds of suitable mechanical properties and _____.

50. If any of the welder essential variables are changed beyond their limits the welder must requalify. (True or False)

51. The purpose of the welder qualification test is to determine the ability of the welder to make sound welds using a previously qualified welding procedure. (True or False)

52. Once a welder is qualified for one of the codes, he can do all types of work without taking any more tests. (True or False)

53. Test coupons that have been subjected to test procedure must not show imperfections of any kind. (True or False)

54. Changing the direction of travel (progression) uphill or downhill for a procedure for which a welder has already qualified requires retesting. (True or False)

55. The highest paid welders are code _____ welders.

56. For a test weld, the least important of the following qualities is _____.

   a. penetration   **b.** appearance   **c.** fusion   **d.** bevel angle

57. Non-destructive testing methods include all but which of the following?

   a. radiographic   **b.** magnetic particle   **c.** ultrasonic   **d.** etching

58. Visual inspection is actually a form of _____ testing and is best applied before, during, and after welding is complete.

59. Radiographic inspection is not used very much because it is expensive and unreliable. (True or False)

60. Ultrasonic inspection should be interpret by a highly experienced inspector. (True or False)

61. Liquid penetrant inspection is used extensively on nonmagnetic materials. (True or False)

Name: _____   Date: _____

62. Magnetic particle inspection can be used on both magnetic and nonmagnetic materials. (True or False)

63. Compressed air is used for hydrostatic testing. (True or False)

64. Removing a cylindrical plug from a pipe weld to inspect the inside surface is called _____.

65. The variable in any welding procedure is the _____.

66. The incomplete filling of the weld root with weld metal is called incomplete _____.

67. Incomplete _____ is the lack of fusion between beads or between the weld metal and base metal.

68. Internal _____ is a condition in which the center of the bead lacks buildup and is below the inside surface of the pipe wall.

69. Poor beveling, preparation, and setup can be as responsible for failure as poor welding technique. (True or False)

70. An insufficient angle of bevel will make it very difficult to secure the proper root penetration and _____.

71. Tack welds are used to hold a joint in proper alignment and to minimize _____.

72. The root opening for a pipe joint should be from 3/32 to _____ of an inch.

73. In tacking up a pipe joint, the easiest way to insure accurate root opening is to use gas-welding _____ as a spacer.

74. Tack welds should never be more than an _____ inch long.

75. Most cross-country pipelines are welded with vertical-_____ travel.

76. Compared to vertical-down welding, vertical-up does all but which of the following?
    a. uses lower current   **b.** uses less electrode per joint
    c. cleans up faster   **d.** melts out gas holes more effectively

77. When welding in the rolled or horizontal fixed position, one layer of weld metal for each _____ inch of pipe wall thickness is recommended.

78. A pipe joint welded in the rolled or horizontal fixed positions requires a _____ inch electrode for the final passes.

79. When pipe in the fixed vertical position is welded on a horizontal plane, the weld metal is deposited in a series of _____ stringer beads.

80. The work connection from the welding machine should be attached as far from the pipe joint as possible. (True or False)

81. Match the following welding excesses to the defects they cause in a finished weld.

1. Arc gap too long	**a.** Excessive spatter, wide bead	**a.** _____
2. Arc gap too short	**b.** Porosity	**b.** _____
3. Slag coverage poor	**c.** Undercutting; high, narrow bead	**c.** _____
4. Travel speed too fast	**d.** Bead pile up	**d.** _____
5. Current too high	**e.** Excessive penetration inside pipe	**e.** _____

82. The hole formation at the tip of the root pass of a pipe butt weld is referred to as the _____.

83. Each pass of a multipass weld on a pipe joint must fuse with the underneath pass and the side _____ of the pipe.

84. To avoid undercut in the weave pass on a horizontal butt joint in the fixed vertical position, pause at the _____ of the weave.

85. The fixed horizontal position for a butt weld on pipe is referred to as the _____ position.

86. The root pass for a butt joint, groove weld in the fixed horizontal pipe axis position, travel up, should be started at the bottom center of the pipe. (True or False)

87. When restarting a weld bead with a new electrode, it is important that the start be made at the end of the previous weld. (True or False)

88. External _____ along the edges of a weld bead is called wagon tracks.

89. Insufficient penetration of a pipe weld may be caused but having all but one of the following? Which one is incorrect?

a. excessive root face

b. the root opening too narrow

c. the electrode diameter too large

d. the welding current too low

90. Internal undercut of a pipe joint will occur in all but which of the following?

a. the root face is too small

b. the root opening is too large

c. the welding current is too high

d. the angle of bevel is too wide

91. The second pass on a pipe weld is often referred to as the _____ pass.

92. A _____ pass is used to build thin sections up to the height of the rest of the weld bead.

93. Because downhill welded beads are thinner than uphill ones, additional _____ passes are required to complete a joint.

94. The cover pass is also referred to as the _____ pass.

95. It is common practice in pipe welding to hand-fabricate as many pipe fittings as possible in the field. (True or False)

72   STUDENT WORKBOOK

CHAPTER 17

Arc Cutting Principles and Arc Cutting Practice: Jobs 17-J1–J7

Please answer the following questions by choosing the letter of the correct answer, circling true or false, or filling in the blanks.

1. The primary advantage of arc cutting is that it can be used on all types of metal. (True or False)

2. Plasma arc cutting is similar in many respects to what common welding process?

3. The fourth state of matter is _____.

4. The process which uses compressed air instead of gravity and the force of the arc to remove molten metal from the work is _____ carbon arc cutting.

5. The air carbon-arc process may be used to prepare plates for welding. (True or False)

6. The air carbon-arc process can be used for fabrication, but it is seldom used for repair and maintenance. (True or False)

7. The depth and contour of the groove when using CAC-A is controlled by all but which one of the following?
    a. electrode angle   **b.** material type
    c. travel speed   **d.** current

8. Both goggles and _____ must be worn during the cutting process for protection from sparks, metal particles, and heat.

9. To cut straight lines accurately, use _____.
    a. straight gouging squares   **b.** magnetic burning tips
    c. rivet-burning tips   **d.** slicing-gauging tips
    e. template or straight edge

10. A portable _____ cutter may be used for trimming and beveling angles and channels.

11. Any alloy with a high hardenability will be in a _____ condition after using the CAC process to cut the metal.

12. What is the recommended current for $\frac{5}{16}''$ diameter carbon arc cutting electrode?

13. Cuts performed in the overhead position with the CAC process are a hazard to the operator because of the molten _____.

14. Oxygen-arc cutting surfaces are rougher than those produced by other arc cutting processes. (True or False)

15. Oxygen-arc cutting cannot be used to cut stainless steel and aluminum. (True or False)

16. A special holder is required for oxygen-arc cutting. (True or False)

17. Fuel gases are not used with arc cutting. (True or False)

18. All but which one of the following is true of carbon arc cutting?
  a. prepares metals for postheat treatment after welding
  b. makes a smooth, accurate cut
  c. produces a very hot arc
  d. allows long service from soft or hard carbon electrodes

19. Arc cutting is superior to oxy-fuel gas cutting because of all but which one of the following?
  a. work is smoother and of better quality
  b. all types of metals can be cut
  c. the heat can be generated solely by electricity
  d. both carbon and metal arcs can be used

20. To be cut by the plasma arc process, a metal must be able to conduct
_____.

21. Regarding the gas used in the plasma arc cutting process, all but one of the following are true. Which one is incorrect?
  a. produces a flame for cutting the work   **b.** shields the operation from the air
  c. maintains an electric arc   **d.** oxidizes the material being cut

22. The high pressure gas flowing through the plasma torch is heated to a temperature of
_____.
  a. 2,700° F   **b.** 50,000° F   **c.** 25,000° F   **d.** 120,000° F   **e.** 150,000° F

23. Stainless steels and nonferrous metals require a _____ gas for plasma arc cutting.

24. Clean cuts can be achieved with the PAC process on most metals up to
_____ inches thick.

25. When cutting carbon steel with the PAC process, superior results are achieved when nitrogen and
_____ are used.

26. Stainless steel and nonferrous alloys are generally cut with mixtures of argon and
_____ when using the PAC process.

27. Magnesium can be cut with the PAC process. (True or False)

28. The _____ arc is an arc between the electrode and the torch tip in the PAC torch.

29. A major concern of _____ with PAC is that it may interfere with telephones, computers, or CNC machines.

30. PAC creates a large heat-affected zone. (True or False)

Name: _____ Date: _____

CHAPTER 18

Gas Tungsten Arc and Plasma Arc Welding Principles

Please answer the following questions by choosing the letter of the correct answer, circling true or false, or filling in the blanks.

1. The GTAW process was originally developed for welding corrosion-resistant and other difficult-to-weld materials such as magnesium, stainless steel, and aluminum. (True or False)

2. Flux cored and metal-cored wires are expansions of what basic welding process?
   _____.

3. In the TIG welding process, the tungsten electrode is consumed. (True or False)

4. It is always necessary to use filler rod when TIG welding. (True or False)

5. The shielding gases most commonly used for TIG welding are argon and
   _____.

6. For the gas metal-arc process all but one of the following are true. Which one is incorrect?
   a. is a nonconsumable electrode process
   b. can operate with a mixture of shield gases
   c. employs one wire as both electrode and filler metal
   d. uses carbon dioxide extensively

7. Because of its high quality, TIG welding is widely used in industry requiring precise welds. (True or False)

8. Gas-shield arc spot-welding can be used to weld both ferrous and nonferrous metals. (True or False)

9. Gas-shielded arc spot-welding has all of the following advantages over resistance spot-welding except one. Which one is incorrect?
   a. access to only the front side of the joint is sufficient
   b. portable equipment can also be used for MIG/MAG and TIG welding
   c. heavy plates can be welded
   d. spatter, smoke, and sparks are minimal
   e. the workpiece suffers little distortion

10. Gas-shielded arc welds are stronger, more ductile, and more resistant to corrosion than those made by the shielded metal-arc process. (True or False)

11. Combinations of dissimilar metals can be welded with the gas-shielded arc process. (True or False)

12. In gas-shielded arc welding, the air in the arc area is displaced by inert gas before the arc is struck. (True or False)

13. The three elements in the air that cause the most weld contamination are oxygen, hydrogen, and _____.

14. Helium and argon must never be mixed for use in TIG welding. (True or False)

15. Both helium and argon are all but which one of following?
    a. chemically inert   **b.** suited to welding thick metal plates
    c. not damaging to tungsten electrodes   **d.** effective mixed together

16. Of the two gases in question 15, which allows faster welding and deeper penetration?

17. Hydrogen may be mixed with helium for TIG welding. (True or False)

18. Nitrogen may be used in TIG welding to weld deoxidized _____.

19. Nitrogen can be used as a backup shield for the TIG welding of stainless steel. (True or False)

20. Helium provides more weld coverage than argon when welding in the vertical and overhead positions. (True or False)

21. _____ indicate the rate of flow of inert gas to the torch.

22. Flowmeters indicate the rate of flow of inert gas to the torch. They are calibrated in which one of the following units?
    a. cubic feet per second   **b.** cubic feet per minute
    c. cubic feet per hour   **d.** cubic feet per inch

23. Too much gas at the end of the tungsten electrode can cause weld _____.

24. The welding machine used for TIG welding is classified as a _____ welder.
    a. constant current and voltage   **b.** variable current and voltage
    c. constant current   **d** variable current

25. The type of operating current directly affects weld _____, contour, and cleaning action.

26. With the GTAW process, _____ frequency allows for the arc to be started without touching the electrode to the work.

27. TIG welding electrodes may be all but which one of the following?
    a. pure tungsten   **b.** thoriated tungsten
    c. zirconium tungsten   **d.** varigated tungsten

28. Gas nozzles used for TIG welding are made of refractory _____ glass, or metal.

29. The TIG torch may be equipped with a gas _____ to prevent turbulence of the gas stream.

30. The _____ feeds both the current and the inert gas to the weld zone for GTAW.

31. The melting point of _____ is higher than that of all other elements except carbon.

32. Pure tungsten electrodes are generally used for TIG welding with a.c. current. (True or False)

33. For thoriated tungsten electrodes, all but one of the following are true. Which one is incorrect?
    a. may be used with a.c. current
    b. are less expensive to use than pure tungsten
    c. do not contaminate the work
    d. facilitate touch starting

34. Tungsten electrodes have a small _____ band to indicate their alloy.

35. The current range for ³⁄₃₂″ diameter pure tungsten when GTAW with ACHF current and argon shielding gas is _____.

36. Grinding of a tungsten electrode should be done on a fine-grit, hard abrasive _____.

37. The length of the tapered end of a correctly ground tungsten should be
    a. ½ times the tungsten diameter
    b. 1 times the tungsten diameter
    c. 0.5 to 2 times the tungsten diameter
    d. 4 times the tungsten diameter

38. The welding end of the tungsten electrode can be pointed or _____.

39. Tap water is generally not recommended for use as a coolant for GTAW torches because of its inherent _____ content.

40. Coolant flowing through the GTAW torch should go directly from the coolant source to the _____.

41. For hot wire welding all but one of the following are true. Which one is incorrect?
    a. is a TIG process
    b. requires preheated filler wire
    c. can be manual or mechanized
    d. concentrates arc heat directly on the weld

42. Welds made with the hot wire process do not require post-weld cleaning. (True or False)

43. TIG hot wire welding is not recommended for aluminum or _____.

44. Plasma welding is done with a.c. current. (True or False)

45. The _____ effect is unique to plasma welding compared to TIG.

46. Plasma arc surfacing is a metal spray process. (True or False)

47. Successful plasma arc surfacing depends upon keeping dilution with the base metal at a minimum. (True or False)

48. Plasma arc surfacing works best on _____ production jobs.

49. The depth of penetration with plasma arc surfacing is precisely controlled to within _____.

50. Because plasma arc surfacing uses _____ and alloys, it is not limited by wire availability.

CHAPTER 19

Gas Tungsten Arc Welding Practice: Jobs 19-J1–J19 (Plate)

Please answer the following questions by choosing the letter of the correct answer, circling true or false, or filling in the blanks.

1. TIG and MIG welding are recommended for aluminum because their _____ gas protects the weld pool.

2. When TIG welding aluminum all but one of the following are true. Which one is incorrect?
   a. there is no glare
   b. there is no smoke
   c. the weld pool is visible
   d. filler metal may be added outside the weld pool

3. The TIG process is preferred for welding aluminum sections up to ½-inch thick. (True or False)

4. The current preferred for TIG welding of aluminum is _____.
   a. DCSP
   b. a.c. with stabilization
   c. DCRP
   d. d.c. with low frequency

5. In addition to any other contamination on an aluminum plate, any _____ film must also be removed.

6. Backup strips for aluminum joints may be made of all but which one of the following?
   a. steel   b. copper   c. stainless steel   d. wrought iron

7. For TIG welding aluminum, all but one of the following are true. Which one is incorrect?
   a. preheating is not always required
   b. distortion is seldom a problem
   c. the weld pool solidifies rapidly
   d. all welding positions may be used

8. When welding aluminum with the TIG process and older style welding machines, rounding the end of the electrode will help the arc _____.

9. For TIG welding carbon and low alloy steels all but one of the following are true. Which one is incorrect?
   a. is easier than welding stainless steels
   b. is used extensively for heavy gauge steels
   c. requires less heat than aluminum from a given electrode size
   d. requires a lighter shade of welding lens than other steels

10. In the TIG welding of carbon steels, best results are obtained with all but which one of the following?
   a. DCEN
   b. argon shielding gas
   c. a balled electrode end
   d. a thoriated electrode

11. The highest quality TIG welds require the use of a _____ control.

12. Stainless steel is widely used for welds in all but one of the following. Which one is incorrect?
   a. for use under high temperatures
   b. to withstand high pressures
   c. with minimal thermal expansion
   d. resistant to corrosion

13. For austenitic stainless steels, all but which one of the following are true?
   a. belong to the 300 series
   b. are hardened by cold working
   c. are highly weldable
   d. may be welded only with direct current

14. For martensitic stainless steels, all but which one of the following are true?
   a. are somewhat brittle
   b. are straight chromium
   c. are hardened by rapid cooling from high temperatures
   d. belong to both 400 and 500 series

15. For ferritic stainless steels, all but which one of the following are true?
   a. are welded with high heat
   b. have high chromium content
   c. belong to the 400 series
   d. can be TIG welded

16. TIG welding of stainless steel eliminates all but which one of the following?
   a. flux
   b. spatter
   c. heavy slag
   d. distortion

17. When making a TIG weld in stainless steel, greater penetration and welding speed can be obtained with _____ current and polarity.

18. Both argon and _____ may be used as the shielding gas for the TIG welding of stainless steel.

19. A mixture of argon and _____ is the equivalent of helium and produces sound welds in austenitic stainless steels.

20. Magnesium is approximately which one of the following?
   a. ⅓ the weight of aluminum
   b. ⅔ the weight of aluminum
   c. 1-½ times the weight of aluminum
   d. the same weight as aluminum

Name: _____ Date: _____

21. For copper, which one of the following is incorrect?

    a. is similar to aluminum for welding purposes

    b. is strongest when cold worked

    c. most commonly uses DCEN for welding

    d. requires higher welding current settings than other metals

22. For TIG welding nickel, which one of the following is incorrect?

    a. eliminates slag entrapment

    b. uses current values in excess of those for carbon steel

    c. attains lower heat for thin materials from ACHF

    d. generally uses DCEN

23. TIG welding is especially suited to welding plate thicknesses up to _____ inch.

24. The backup for TIG welding may be _____ painted on the weld underside.

25. A gas recommended for backing for stainless steels is _____.

26. A ³⁄₁₆-inch tungsten electrode is the largest size available. (True or False)

27. A dirty tungsten electrode indicates that during a previous use all but which one of the following occurred?

    a. the gas was shut off before the electrode was cooled

    b. there was an air leak in the gas supply system

    c. the tip was contaminated by touching the weld pool

    d. the welding tip was oxidized

28. When TIG welding with a current of 75 to 200 amperes, a welding operator should wear shaded lenses of glass number _____.

    a. 6   **b.** 10   **c.** 12   **d.** 14

29. For TIG welding with a.c. high frequency current, the electrode does not have to touch the work to start the arc. (True or False)

30. Which of the following is not one of the three arc starting methods generally used for TIG welding?

    a. high frequency starting   **b.** scratch starting

    c. low-voltage touch starting   **d.** lift arc starting

31. During TIG welding, arc wandering may be caused by all but which one of the following?

    a. current set too high   **b.** a contaminated electrode

    c. magnetic effects   **d.** drafts in the work area

32. During TIG welding, it is important to maintain a _____ arc.

33. The extension of the tungsten electrode beyond the gas cup is governed by the shape of the object and the type of _____.

34. Sharpen the electrode end to a fine point in all but which one of the following?

    a. welding light-gauge carbon steel   **b.** welding aluminum

    c. concentrating heat on a limited area   **d.** welding stainless steel

35. If the addition of filler rod is necessary, choose a rod size greater than the thickness of the metal being welded. (True or False)

36. During TIG welding, the longer the extension of the electrode beyond the cup, the less effective the gas shielding. (True or False)

37. To finish a TIG weld, the arc should be broken abruptly. (True of False)

38. A gas _____ can be used where a long electrode extension is required to aid in getting proper gas coverage.

39. Of the metals most commonly TIG welded, the easiest to weld is _____.

    a. aluminum   **b.** carbon steel

    c. low alloy steel   **d.** stainless steel

40. During TIG welding, an erratic arc may occur in all but which one of the following?

    a. the electrode diameter is too large

    b. the arc is too long

    c. the joint is too wide

    d. the base metal is dirty

41. During TIG welding, the hot filler rod should be within the shielding gas area to prevent excessive oxidation. (True or False)

42. To control penetration and weld contour during TIG welding, use an arc length _____.

    a. equal to the electrode diameter

    b. smaller than the electrode diameter

    c. a little longer than the width of the joint

    d. a little shorter than the width of the joint

43. For TIG welding, always use the largest gas cup possible for a given weld. (True or False)

44. Alternating the addition of filler metal to the weld pool with alternated torch movement makes a better appearing bead. (True or False)

45. The easiest joint to weld with the TIG process is the _____.

    a. butt   **b.** T   **c.** lap   **d.** corner   **e.** edge

46. Filler rods must be added to TIG welded _____.

    a. edge joints   **b.** lap joints   **c.** T-joints   **d.** butt joints

47. The center of the weld pool for TIG-welded lap joints is called the _____.

48. Multipass fillet welding with the TIG process is generally necessary when the material thickness is more than _____.

    a. $1/8''$ **b.** $3/16''$ **c.** $5/32''$ **d.** $1/4''$

49. The single-V butt joint is used when complete penetration is required on material thicknesses ranging between _____.

50. In a stainless steel groove weld, a dark purplish-blue bead with hardly noticeable ripples indicates excessive _____.

CHAPTER 20

Gas Tungsten Arc Welding Practice: Jobs 20-J1–J17 (Pipe)

Please answer the following questions by choosing the letter of the correct answer, circling true or false, or filling in the blanks.

1. The TIG welding process can be used for welding ferrous and nonferrous piping materials. (True or False)

2. Aluminum is preferred for applications requiring all but which one of the following?
   a. light weight
   b. corrosion resistance
   c. temperatures above 750° F
   d. nontoxic materials for food processing

3. Metals especially resistant to corrosive chemicals include all but which one of the following?
   a. chromium **b.** stainless steel **c.** copper nickel

4. For welding carbon steel pipe, the TIG process is used as a common choice for the _____ pass only.

5. The process of passing argon or helium through the pipe to shield the weld pool from the inside is called _____.

6. The inside of a pipe may be protected by painting it with an appropriate _____.

7. The formation of carbon monoxide or carbon dioxide during steel welding will cause _____ in the weld.

8. Porosity may result from the presence on a weld joint of all but which one of the following?
   a. rust **b.** flux **c.** oil **d.** grease **e.** moisture

9. Porosity caused by inadequate gas shielding may be the result of all but which one of the following?
   a. gas flow that is too slow
   b. an electrode that is too large
   c. an arc that is too long
   d. air currents in the work area

10. When TIG welding pipe, the presence of a notch in the weld pool indicates the absence of penetration on the inside of the pipe. (True or False)

11. Add filler rod to the weld pool to correct _____ penetration in the root pass of a TIG weld on pipe.

12. Walking the cup is a method that allows for more control on the _____ pass.

13. Control of _____ is the most important factor in successful root-pass welding.

14. To terminate a weld bead, direct the weld pool off to the side of the joint. (True or False)

15. Filler passes are used to fill the pipe joint to within _____ inch of the top surface of the pipe.

16. Filler passes may be made using the stringer bead technique or the weave bead technique. (True or False)

17. Finished cover passes should include all but which one of the following?

    a. be ³⁄₃₂″ wider than the beveled edges on each side of the joint

    b. have no overlap

    c. be slightly convex

    d. have a crown about ¹⁄₁₆″ above the pipe surface

18. Pipe with axis in the vertical orientation should be welded with _____ beads.

19. Pipe with axis in the vertical orientation requires special techniques to compensate for _____ of the weld pool caused by gravity.

20. Pipe with axis in the horizontal fixed orientation enables a welder to practice three welding positions. (True or False)

21. Pipe with axis in the horizontal fixed orientation, the weld bead should be started at the bottom center. (True or False)

22. Downhill travel does not produce satisfactory results when pipe is being TIG welded. (True or False)

23. When welding pipe with the TIG process, the second pass must be applied with great care because the root pass is likely to be _____.

24. Soundness tests for TIG pipe welds are similar to those for plate. (True or False)

25. Lateral joints may also be referred to as _____ joints.

CHAPTER 21

Gas Metal Arc and Flux Cored Arc Welding Principles

Please answer the following questions by choosing the letter of the correct answer, circling true or false, or filling in the blanks.

1. Gas metal-arc welding is a continuous _____ process.

2. The constant current welding machine is used for MIG/MAG welding. (True or False)

3. The use of the MIG/MAG process was extended by the introduction as a shield gas of _____.

   a. carbon dioxide   **b.** argon   **c.** hydrogen

   d. helium   **e.** oxygen

4. Some forms of MIG/MAG welding involve the use of flux. (True or False)

5. Match the types of open arc transfer to their best descriptions.

1. Spray	**a.** Current alternates between high and low ranges.	**a.** _____
2. Globular	**b.** Transfer occurs below the surface of the base metal.	**b.** _____
3. Pulsed arc	**c.** Molten ball twice the diameter of the filler wire is transferred.	**c.** _____
4. Buried arc	**d.** Fine droplets of metal move rapidly from electrode to workpiece.	**d.** _____

6. In short-circuiting arc transfer, no metal is transferred across the welding arc. (True or False)

7. Use only argon and helium mixtures as shielding gases for short circuit transfers. (True or False)

8. Instantaneous, flawless arc starts are characteristic of constant _____ machines.

9. In MIG/MAG welding, the amount of current is determined by the _____ at which the electrode wire is fed to the work.

10. The constant voltage power source is designed to keep the arc length constant by varying _____ rate.

11. During MIG/MAG welding, arc voltage is higher than open circuit voltage. (True or False)

12. The controls commonly found on constant voltage machines regulate all but which one of the following?

    a. voltage   **b.** current   **c.** slope   **d.** inductance

13. _____ is the force that causes current to flow but does not flow itself.

14. Most MIG/MAG welding is done using _____.

    a. a.c. straight polarity

    b. d.c.e.n polarity

    c. d.c.e.p. polarity

    d. a.c. reverse polarity

15. Most wire feeders _____ the wire through the cable to the torch.

16. Wire feeders that pull wire through the cable to the torch use wire that is all but which one of the following?

    a. hard   **b.** soft   **c.** aluminum   **d.** manganese

17. Shielding gas nozzles must be large enough to allow proper gas flow yet small enough to allow access to the _____.

18. All MIG/MAG welding guns have a _____ switch that turns on controls for wire feed, shielding gas, and current-contactor.

19. The shielding gas system supplies and controls the flow of gas which shields the arc area from the surrounding _____.

20. Gas from the cylinder is controlled by a pressure-reducing regulator, solenoid control valves, and a _____.

21. To provide a good gas shield, _____ flow must be established.

22. A gas flow rate that is too high causes _____.

23. Of the following, the least dense gas is _____.

    a. air   **b.** carbon dioxide   **c.** helium   **d.** argon

24. Denser gases require higher flow rates for adequate shielding than do lighter gases. (True or False)

25. Such inert gases as xenon, krypton, and neon are seldom used for MIG/MAG welding. (True or False)

26. To stabilize the welding arc, use _____ alone or mixed with another shielding gas.

27. When used under appropriate conditions, argon shielding produces a narrow weld with deep penetration at the center. (True or False)

28. Argon is preferred for welding heavy gauge metals. (True or False)

29. Argon should never be used for out-of-position welding. (True or False)

30. Helium shielding requires a _____ gas flow rate than argon shielding.

31. Helium shielding produces a _____ slightly less penetrating weld than argon shielding.

32. Helium is seldom used with automatic and mechanized processes. (True or False)

33. Carbon dioxide gas is all but which one of the following?
   a. an inert gas
   b. composed of carbon and oxygen
   c. a contributor to weld spatter
   d. used for deep penetration
   e. known to cause an unstable arc

34. Welding with carbon dioxide produces poisonous gas. (True or False)

35. The presence of even a little moisture in carbon dioxide gas affects the quality of the weld. (True or False)

36. Shielding gases that are inert may be mixed with shielding gases that are not. (True or False)

37. Materials that require filler wire of composition different than the base metal are all but which one of the following?
   a. copper alloys
   b. zinc alloys
   c. magnesium alloys
   d. high strength aluminum
   e. high strength steel alloys

38. Match the welds in Fig. 21-1 to the shielding gases used to make them.

1. Helium/Argon   **a.** _____
2. Argon/CO$_2$ **b.** _____
3. Argon   **c.** _____
4. CO$_2$ **d.** _____
5. Helium   **e.** _____
6. Argon/Oxygen   **f.** _____

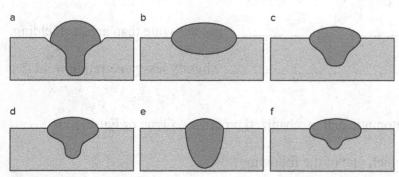

Fig. 21-1.

39. Filler wires as large as ¼ inch in diameter are available for MIG welding. (True or False)

40. To prevent weld porosity, deoxidizers are used as scavenging agents. (True or False)

41. The T in the AWS classification for flux-cored filler wires stands for
_____.

42. The flux is the primary method of carrying the _____ and deoxidizing elements to the weld pool in a flux cored electrode.

Name: _____ Date: _____

CHAPTER 22

Gas Metal Arc Welding Practice with Solid and Metal Core Wire: Jobs 22-J1–J23 (Plate)

Please answer the following questions by choosing the letter of the correct answer, circling true or false, or filling in the blanks.

1. _____ current is generally used for GMAW.

2. For GMAW, _____ or carbon dioxide may be added to the shielding gas to reduce porosity and improve arc stability and weld appearance.

3. Argon shielding gas gives better GMAW coverage than helium because it is _____ than air.

4. The use of carbon dioxide for GMAW has all of the following advantages except which one?
   a. low cost   **b.** low density   **c.** low flow rates   **d.** less burn back

5. Pure argon is the preferred shielding gas for all but which one of the following?
   a. titanium   **b.** magnesium   **c.** nickel   **d.** stainless steel

6. Add _____ to argon to increase penetration.

7. Joint designs like those used for other welding processes can be used for GMAW with a reduction in costs. (True or False)

8. Compared to shielded metal-arc welding, GMAW is all but which one of the following?
   a. is more penetrating
   b. allows wider groove-plate openings
   c. is faster
   d. requires less weld metal

9. The constant voltage welding machine used for gas metal arc welding provides for the self-adjustment of the arc length. (True or False)

10. During GMAW the drag angle technique produces deeper penetration than the push angle technique. (True or False)

11. When welding a T-joint with GMAW, the _____ angle can be used to prevent undercut.

12. In GMAW, if the current is too high, penetration may be too deep. (True or False)

13. In GMAW, the electrode wire-feed speed determines the welding current. (True or False)

14. For GMAW, the operating variables are adjusted on the basis of all but which one of the following?

   a. the type of material being welded   **b.** the thickness of the material

   c. the position of welding   **d.** type of welding machine used

15. If GMAW wire-feed speeds are excessive, _____ or stubbing results because there is not enough current to melt the wire fast enough.

16. If a given GMAW travel speed and wire-feed speed are not providing sufficient weld metal, which one of the following should occur.

   a. increase both   **b.** increase travel or decrease wire feed

   c. decrease travel or increase wire feed   **d.** decrease both

17. Match the following GMAW discontinuities to the combination of procedures most likely to correct them.

1. Incomplete penetration	**a.** Use shield gas properly; clean and maintain equipment; reduce current	**a.** _____
2. Excessive penetration	**b.** Reduce current or increase travel speed	**b.** _____
3. Whiskers	**c.** Control expansion and contraction	**c.** _____
4. Voids	**d.** Weave welding gun; hesitate at each side of the joint	**d.** _____
5. Incomplete fusion	**e.** Increase current; reduce stickout distance	**e.** _____
6. Porosity	**f.** Use compatible filler and base metals; allow for expansion and contraction	**f.** _____
7. Warpage	**g.** Handle torch properly; use appropriate shield gas; maintain proper travel speed, current, arc voltage, and stickout	**g.** _____
8. Cracking	**h.** Reduce travel and wire-feed speeds; increase stickout; weave gun	**h.** _____
9. Excessive spatter	**i.** Cover all joint areas with arc; reduce pool size; lead pool with electrode; check current values	**i.** _____
10. Irregular weld shape	**j.** Reduce current; shorten and stabilize arc; clean nozzle; use appropriate shielding gas	**j.** _____
11. Undercutting	**k.** Fill all passes; increase arc voltage and travel speed following defective passes	**k.** _____

18. For the radiation produced by GMAW, all but which one following is true?

   a. is the highest produced by any welding process

   b. increases with the use of argon

   c. disintegrates cotton clothing

   d. increases spatter from hot metal and slag

19. For maximum eye protection during GMAW, all but which one of the following is correct?

    a. a lens that is too dark should be avoided to prevent eye strain

    b. always shield the eyes when less than 20 feet from the arc

    c. wear safety glasses under the welding helmet

    d. use a heavier lens shade for indoor welding

20. Special ventilation precautions should be taken when GMAW under all but which one of the following?

    a. metals that emit toxic fumes

    b. near degreasers

    c. using carbon dioxide shielding gas

    d. using nitrogen shielding gas

    e. in the presence of any amount of ozone

21. Electrical hazards are not quite as great with GMAW as other processes. (True or False)

22. Whenever possible, GMAW should be done in the _____ position to increase penetration and deposition rates.

23. MIG/MAG welding requires the use of _____ current welding machines.

24. The current value on a MIG welding machine can be checked on the machine ammeter when the welder is not welding. (True or False)

25. Arc voltage controls all but which one of the following?

    a. penetration   **b.** bead contour   **c.** undercutting

    d. bead size   **e.** porosity

26. GMAW is more expensive than other welding processes. (True or False)

27. Fillet welds made by the GMAW process may be smaller than those made with stick electrodes. (True or False)

28. The GMAW process is generally used for welding _____ gauges of aluminum.

29. Minimize arc blow by welding away from the work connection. (True or False)

30. With GMAW, the current level is adjusted using the _____ knob.

31. Aluminum _____ joints perform better under fatigue loading than other kinds of joints.

32. Porosity in aluminum GMAW welds is commonly caused by all but which one of the following?

    a. hydrogen in the weld area   **b.** a leaky gun

    c. improper joint preparation   **d.** insufficient shielding gas

33. GMAW welding stainless steel does all but which one of the following?

    a. reduces cleaning time   **b.** makes the weld pool visible

    c. allows continuous wire feed   **d.** uses only the fast, spray arc technique

34. All but which one of the following metals are readily welded by the GMAW process?

    a. titanium   **b.** Monel®   **c.** electrolytic copper

    d. magnesium   **e.** Inconel®

35. Argon is recommended for GMAW of magnesium because of its excellent
_____ action.

36. Lead, phosphorus, and sulfur improve nickel alloys resistance to cracking when heated. (True or False)

37. All but which one of the following is true of titanium?

    a. is the only element that burns in nitrogen

    b. melts at a low temperature

    c. is used in electrode coatings

    d. is used as a stabilizer in stainless steel

38. All but which one of the following is true of titanium?

    a. remains solid up to 3,500° F.   **b.** scratches glass

    c. is used in hard-facing material   **d.** is usually used in the pure state

39. Special shielding precautions are necessary for gas metal arc welding titanium and zirconium. (True or False)

40. _____ is the preferred shielding gas for welding aluminum plate up to 1 inch.

Name: _____ Date: _____

CHAPTER 23

Flux Cored Arc Welding Practice (Plate), Submerged Arc Welding, and Related Processes: FCAW-G Jobs 23-J1–J11, FCAW-S Jobs 23-J1–J12, SAW Job 23-J1

Please answer the following questions by choosing the letter of the correct answer, circling true or false, or filling in the blanks.

1. Flux-cored arc welding increases welding speeds and _____ rates.

2. Flux-cored arc may be gas or _____ shielded.

3. Both types of flux cored arc welding can produce welds of the highest
   _____.

4. FCAW-G electrodes have deposition efficiencies up to 90 percent; where FCAW-S electrodes have less deposition efficiencies, up to 87 percent. (True or False)

5. Direct current electrode _____ is most commonly used for FCAW-S electrodes.

6. FCAW-G electrodes are used with such shielding gases as all but which of the following?
   a. carbon dioxide **b.** nitrogen   **c.** argon plus carbon dioxide

7. All but which of the following flux-cored filler electrodes may be used with external shielding gas?
   a. E70T-1   **b.** E70T-2   **c.** E70T-3   **d.** E71T-9

8. All but one of the following flux-cored filler electrodes require no external shielding gas. Which one is incorrect?
   a. E70T-3   **b.** E71T-7   **c.** E70T-10   **d.** E70T-12

9. Because flux-cored electrodes carry an AWS Classification number, it is acceptable to switch between the various manufacturers. (True or False)

10. M type electrodes require _____ % argon/balance CO_2; if M is not present CO_2 is to be used.

11. It may be necessary to _____ gas-flow rates when flux-cored electrode welding is done outside.

12. When welding with flux-cored arc welding process, a push (forehand) nozzle angle does all but which of the following?
    a. directs gas shield ahead of weld puddle   **b.** plays the arc stream on cold base metal
    c. produces a narrow weld bead   **d.** reduces burn-through on thin plate.

13. Gas-shielded, flux-cored arc welding is usually done with a minimum extension in inches of
_____.

    a. ⅛″   **b.** ¼″   **c.** ½″   **d.** 1″

14. Self-shielded, flux-cored arc welding is usually done with a minimum extension in inches of
_____.

    a. ⅛″   **b.** ¼″   **c.** ½″   **d.** 1″

15. Electrode extension that is too long results in all but which of the following?

    a. excessive spatter   **b.** irregular arc action

    c. loss of gas shielding   **d.** excessive penetration

16. Gas-shielded, flux-cored arc welding deposits are lower in hydrogen than those made with low hydrogen stick electrodes. (True or False)

17. During gas-shielded, flux-cored arc welding, all but one of the following are true. Which one is incorrect?

    a. too long an arc causes spatter

    b. cleaning the plate reduces undercut

    c. penetration decreases as travel speed increases

    d. drag welding produces deeper penetration

18. When welding with gas-shielded, flux-cored arc electrodes, travel speed that is too fast for the current produces all but one of the following weld characteristics. Which one is incorrect?

    a. slag entrapment   **b.** undersized bead

    c. shallow penetration   **d.** undercutting

19. The equipment needed for self-shielded, flux-cored arc welding includes all but one of the following?

    a. constant voltage power source

    b. continuous wire-feed mechanism

    c. auxiliary gas shielding

20. The self-shielded, flux-cored arc contains all the necessary ingredients for shielding, deoxidizing,
_____ and alloying materials.

21. The self-shielded, flux-cored wire process offers all but which of the following advantages?

    a. all-position welding

    b. faster, more flexible installation

    c. heavier, more stable welding guns

    d. uninterrupted wire feeding

22. Constant voltage, d.c, power source are used for welding FCAW-S process and electrodes. These machines can be all but which of the following types?

    a. transformer rectifier type

    b. inverter type

    c. semiautomatic type

    d. engine-driven generator type

23. Only about ½ of the total extension length is visible beyond the end of the nozzle on the welding gun. (True or False)

24. Increasing the arc voltage during self-shielded, flux-cored arc welding need not affect the other three welding variables. (True or False)

25. Submerged arc welding takes place beneath a blanket of granular, fusible _____ which completely hides the arc.

26. For submerged arc welding, which of the following is incorrect?

    a. the filler wire lightly touches the workpiece

    b. there is no spatter, smoke, or sparks

    c. goggles are the only eye protective equipment needed

    d. the process may be automatic or semiautomatic

27. Some smoke and fumes are created that require proper _____.

28. Submerged arc welds have all but which of the following characteristics?

    a. high impact strength   b. uniformity   c. density
    d. corrosion resistance   e. high nitrogen content

29. Submerged arc welding must be done in the _____ position.

30. For alloy pipe welding with the submerged arc process, the alloying elements should be mixed with the flux. (True or False)

31. The power source for submerged arc welding may be AC, DCEP, or DCEN. (True or False)

32. _____ provides the best control of bead shape and maximum penetration.

33. For constant voltage welding machines, which one of the following is not correct?

    a. use constant-feed speed wire feeder

    b. are preferred for small diameter filler wire

    c. are recommended for hard-surfacing

    d. are preferred for high-speed welding of thin materials

CHAPTER 23   FCAW Practice (Plate), SAW, and Related Processes   **99**

34. For the fluxes used for submerged arc welding, which of the following is not true?

    a. carry high welding currents

    b. clean the base metal

    c. add no new properties to the weld metal

    d. provide maximum resistance to cracking

35. The fluxes used with the submerged arc welding process contain all but which of the following?

    a. silicon   **b.** chromium   **c.** manganese

    d. aluminum titanium   **e.** deoxidizers

36. The largest diameter filler wire used with the submerged arc welding process is ¼ inch. (True or False)

37. DCEP is recommended for tandem arc welding applications. (True or False)

38. Heavy, deep-groove submerged arc welding is often done slightly downhill. (True or False)

39. Joints with gaps greater than ¹⁄₁₆″ may be filled with SMAW, GMAW, or FCAW process. (True or False)

40. When welding long seams on a tank, eliminate cracking by tack welding steel _____ at each seam end.

41. Excessive flux produces a narrow hump bead. (True or False)

42. For submerged arc welding, a long extension reduces costs by increasing the deposition rate and the welding speed. (True or False)

43. During submerged arc welding at a fixed current setting, increasing the electrode size affects the _____ of penetration.

44. For multiple arcs, which of the following is not correct?

    a. increase meltoff   **b.** increase speed

    c. use one power source   **d.** reduce deposition rates

45. The SAW process is often used to _____ carbon steel with stainless steel.

46. Semiautomatic submerged arc welding is a _____ wire process recommended for irregular shapes.

47. A submerged arc process developed especially for single-pass welding of thick vertical plates is _____ welding.

48. For semiautomatic submerged arc welding, which of the following is not correct?

    a. requires no plate edge preparation   **b.** is a manual process

    c. smooth, spatter free bead   **d.** is a high-heat process

49. Consumable-guide electroslag welding is not a true submerged arc process. (True or False)

50. EGW can be done with solid metal cored or flux cored electrodes. (True or False)

51. During electrogas welding, the weld pool is protected by shielding gas. (True or False)

52. Electrogas welding uses direct current electrode positive. (True or False)

53. The selection of the welding process depends on the proper evaluation of each job. (True or False)

54. Automatic welding processes are ideal for _____.

    a. short welds   **b.** complicated shapes

    c. repetitive, fixtured jobs   **d.** heavier construction

CHAPTER 24

Gas Metal Arc Welding Practice: Jobs 24-J1–J15 (Pipe)

Please answer the following questions by choosing the letter of the correct answer, circling true or false, or filling in the blanks.

1. For MIG/MAG welding pipe, which one of the following is not true?
   a. is often faster than other processes
   b. increases cleaning time
   c. may not require backup rings
   d. reduces warpage and distortion

2. A d.c. constant current welding machine is used for the MIG/MAG welding of pipe. (True or False)

3. For MIG/MAG welding, a quiet power source that draws current only during welding is
   _____.
   a. a motor-driven generator
   b. a d.c. rectifier
   c. an engine-driven generator
   d. an a.c.- d.c. transformer-rectifier

4. Power supplies for MIG/MAG welding should have all but which one of the following?
   a. variable slope control
   b. an inverter
   c. a hot start feature
   d. a special shut-off control

5. A gas flow rate of _____ cubic feet per hour is adequate for most indoor welding applications.

6. For spray transfer GMAW, the minimum argon content is _____.
   a. 70%   b. 80%   c. 90%   d. 100%

7. A tri-mix gas for GMAW of stainless steel would consist of all but which of the following gases?
   a. helium   b. nitrogen   c. argon   d. CO_2

8. GMAW is considered to be a low hydrogen process. (True or False)

9. For best results, pipe should be welded with E70S-3 filler wire with a diameter of
   _____ inch.

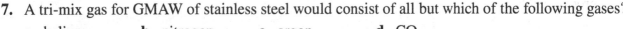

10. Using a smaller filler wire for MIG/MAG welding at a given current level requires decreasing the wire feed speed. (True or False)

11. Match the common MIG welding discontinuities to the appropriate remedies.

1. Convex bead	**a.** Clean weld thoroughly
2. Scattered porosity	**b.** Increase root opening and voltage
3. Cold lapping	**c.** Speed up travel; decrease root opening
4. Lack of root fusion	**d.** Decrease voltage, increase root opening, decrease root face
5. Suckback in overhead position	**e.** Lower wire-feed speed; increase forward speed; keep arc ahead of puddle

12. To remove tack defects and to insure good fusion to tack welds for MIG/MAG weldng, both ends of all tacks should be _____.

13. If a backing ring is used for MIG/MAG welding, a root opening of at least _____ inch is necessary.

14. The _____ joint is the most commonly used pipe joint in welded pipe systems.

15. A 37½-degree bevel and uphill travel are used for MIG/MAG welding _____ piping.

16. Cross-country and _____ piping have a 30-degree bevel and are welded downhill.

17. Over _____ percent of all steel piping installations are welded.

18. The bell-hole position is another name for _____.
  a. horizontal roll pipe axis
  b. vertical fixed pipe axis
  c. horizontal fixed pipe axis
  d. vertical roll pipe axis

19. It is important to keep the arc _____ of the weld pool when doing filler and cover downhill

20. The _____ joint requires the welding of Tee and lap joints with fillet and groove welds.

21. The downhill welding technique is employed in cross-country transmission line piping, and the uphill welding technique is employed in pressure piping. (True or False)

Name: _____ Date: _____

CHAPTER 25

High Energy Beams and Related Welding and Cutting Process Principles

Please answer the following questions by choosing the letter of the correct answer, circling true or false, or filling in the blanks.

1. High energy beams are concentrated heat sources that have measured as high as _____ watts/square inch.

2. All normal type joints can be welded with high energy beam processes. (True or False)

3. High energy beams generally produce very narrow welds with very deep _____.

4. Complete joint penetration (CJP) welds using the keyhole technique require joint tolerances in the range of _____.
   a. +/−0.0001″
   b. +/−0.001″
   c. +/−0.010″
   d. +/−0.100″

5. Which high-energy process is accomplished by the use of a concentrated stream of high-velocity electrons formed into a beam?

6. Laser beam welding (LBW) is generally done with an _____ gas to shield the weld pool.

7. Lasers can cut non-metals, such as plastic wood or cloth. (True or False)

8. _____ are unwanted molten metal flying out of the cut, interrupting the cut path.

9. When laser beam cutting, an _____ gas is used to help improve combustion and physically blow metal from the kerf.

10. The assist gas recommended for laser beam cutting of copper is _____.
    a. air   b. oxygen   c. nitrogen   d. argon

11. Which one of the following is not an advantage of a laser beam?
    a. very low heat input
    b. noncontact process
    c. very few metals can be processed
    d. very small welds and holes can be made

12. Water jet cutting uses a high-velocity jet of water at pressures of up to
_____ psi.

13. Which of the following materials is not used for the orifice of a water jet cutting nozzle?

    a. jade   **b.** diamond   **c.** sapphire   **d.** ruby

14. Water jet cutting produces narrow kerfs of _____ inch.

15. How fast can abrasive water jet cut ¾″ armor plate steel?

16. The welding process where a probe or tap with a diameter of 0.20″ to 0.24″ is rotated between the square groove faying edges on a butt joint is _____.

    a. laser beam   **b.** electron beam

    c. forge   **d.** friction stir

17. A solid state process that uses a controlled detonation to impact two work pieces at a very high velocity is _____.

18. A solid state process that uses the heat produced by compressive forces generated by materials rotating together is _____.

19. Laser assisted arc welding is a hybrid process using a laser in combination with _____.

    a. SMAW   **b.** SAW   **c.** FCAW   **d.** GMAW

20. The American Welding Society has defined and described over 110 various joining and cutting processes and process variations. (True or False)

CHAPTER 26
General Equipment for Welding Shops

Please answer the following questions by choosing the letter of the correct answer, circling true or false, or filling in the blanks.

1. In shop areas designed specifically for welding, permanent _____ are erected to protect other welders.

2. The use of positioners to enable welding in the flat position has done all but which of the following?
   a. decreased production
   b. reduced costs
   c. improved quality
   d. promoted safety in both production and repair welding

3. Turning rolls are used to weld circumferential seams in _____ position.

4. There are two main types of turning rolls: those with steel or rubber-tired rolls and those with roller chain slings. (True or False)

5. A weld _____ has three movable jaws similar to the ones on a lathe chuck.

6. A _____ enables arc welding units to move vertically and horizontally as well as to rotate a full 360°.

7. A turntable is used for overhead welding. (True or False)

8. There are four basic types of seamers. (True or False)

9. A weld _____ raises or lowers the welder himself along a vertical surface.

10. There are a number of clamps and holding devices that employ magnetic attraction. (True or False)

11. For precision linear travel, a side beam carriage is an effective tool. (True or False)

12. There are many advantages to hydroforming, including all but which of the following?
    a. reduce tooling cost   b. increase part weight
    c. reduce the number of weld joints   d. increase part stiffness

13. An orbital welding machine is used to make fillet welds in all positions. (True or False)

14. The C clamps and wedges important to carpentry are merely excess baggage for welders. (True or False)

15. To be sure that holes in hot metal do not lose their shape, reinforce them with round _____ sticks.

16. Heat-treating ovens cannot be fired by electricity. (True or False)

17. Cleaning a surface for welding and removing scale, slag, and rust after welding, is done by _____.

18. Materials used as abrasives in sandblasting equipment include all of the following except one. Which one is incorrect?

    a. aluminum oxide   **b.** metal grit

    c. lead filings   **d.** walnut shells

    e. corn cobs

19. Using heat and pressure to weld two lapped pieces is _____ welding.

20. In modern welding shops, hand tools have been replaced by _____ ones.

21. A versatile brake that can form four sides and a bottom from one flat metal sheet is a _____.

    a. box and pan brake   **b.** hand brake

    c. roll   **d.** power press brake

22. An indispensable piece of equipment for bending radii and angles on various shapes is a _____ bender.

23. Power squaring shears may be used on metal plate not more than ½-inch thick. (True or False)

24. Match the tools in Fig. 26-1 to the appropriate names.

    1. Weld gauges   **a.** _____

    2. Micrometer   **b.** _____

    3. Pliers   **c.** _____

    4. Cold chisel   **d.** _____

    5. Identification stamp   **e.** _____

    6. Center head   **f.** _____

    7. Combination square and scale **g.** _____

    8. Hand hammer   **h.** _____

    9. Scribe and center punch   **i.** _____

    10. Bevel protractor   **j.** _____

    11. Flexible scale   **k.** _____

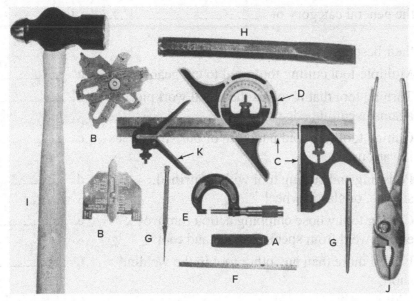

Fig. 26-1.

25. Of the four types of portable power tools, universal electric ones are most common. (True or False)

26. For universal electric power tools, all of the following are true except which one?
    a. use the cheapest power available
    b. are ideal for continuous operation
    c. require expensive maintenance
    d. meet the demands of field work

27. Compressed air is required for the operation of _____ power tools.

28. Pneumatic tools are suitable for continuous operation. (True or False)

29. Hydraulic power tools are operated by hydraulic _____.

30. Match the following power tools to their best descriptions.

1. Electric hammers	a. Produce a smooth cutting edge without distorting the body of the metal	a. _____
2. Grinders	b. Used for grinding, buffing, and finishing aircraft, missile, and industrial operations	b. _____
3. Shears and nibblers	c. Used for chipping, peening, channeling, and masonry drilling	c. _____
4. Magnetic-base drill press	d. Used for economic, precision edge preparation	d. _____
5. Beveling machine	e. Attached to steel surfaces for difficult drilling jobs	e. _____
6. Weld shaver	f. Abrasive surfacing tools	f. _____

CHAPTER 26   General Equipment for Welding Shops   **109**

31. Tools that cut metal fit into the general category of _____ tools.

32. Match the machine tools to their best descriptions.

1. Engine lathe	**a.** Multiple-tool cutting tool used to cut gear teeth	**a.** _____
2. Shaper	**b.** Turning tool that revolves the metal workpiece against a cutting edge	**b.** _____
3. Milling machine	**c.** Cutting tool for rapid puncture of burr-free holes in various sizes	**c.** _____
4. Pedestal grinder	**d.** Polishing and cutting tool with a vitrified, silicate, or elastic wheel	**d.** _____
5. Power punch	**e.** Cutting tool whose chipping action removes excess weld from specimen face and root	**e.** _____
6. Metal cutting band saw	**f.** Is used more than any other tool in the welding shop	**f.** _____

CHAPTER 27

Automatic and Robotic Arc Welding Equipment

Please answer the following questions by choosing the letter of the correct answer, circling true or false, or filling in the blanks.

1. In the automatic and robotic method of applications, the welders or operators are required to control the key factors influencing the weld. (True or False)

2. In certain situations, it is more advantageous to move the arc and thus the weld pool than it is to require the motion control device to change the torch or gun angle. (True or False)

3. Magnetic arc controls are typically used with all but which one of the following processes?
    a. GMAW   **b.** GTAW   **c.** SAW   **d.** PAW

4. The devices in question 3 provide control over all but which one of the following?
    a. heat distribution   **b.** excessive porosity
    c. incomplete fusion   **d.** a wandering arc

5. Sophisticated mechanical oscillators are available, such as the _____ controlled pendulum.

6. Arc monitoring does not require the welder or operator to be skilled or trained. (True or False)

7. There are many variables that must be dealt with that may adversely affect the completed weldment. (True or False)

8. There are four basic control elements that must be dealt with. Which one of the following is not one of the four?
    a. welding
    b. manipulation of the input variables
    c. what is happening with the disturbing input variables
    d. mission control

9. The manipulated input variables are those that directly affect the _____ response variables.

10. Sensing devices are generally transducers that convert energy from one form to another. (True or False)

11. _____ of the various types are of no value unless their data can be monitored and used to control the deviation that is taking place.

12. Control over a welding operation for automatic or robotic applications is very involved; the fundamental goal is to deposit a satisfactory weld. (True or False)

13. Having good control over sequence of operation of the weld or "weld sequence" is

_____.

14. Microprocessor-based controllers allow digital readouts and accurate setting of the weld sequence. (True or False)

15. Robotic arc welding systems are very flexible. (True or False)

16. The two main types of robot arms are the _____ and the rectilinear.

17. In 1997 NASA's robot lander, Pioneer, crawled across the surface of

_____.

18. The pool of professional welders is dwindling; in fact, _____ percent of the current welding workforce will be retiring in the next decade. This will leave a shortfall of skilled, knowledgeable welders.

19. Robots do not always have the ability to return to the exact same position each time. (True or False)

20. Robots are rated by how many inches a minute an axis can move in a second. (True or False)

21. Occasionally a robot may have to be programmed for up to _____ axes of motion.

22. There are _____ different levels of recommended robotic training for operations personnel.

23. The AWS QC19 is a Standard for _____.
   a. welder operations
   b. qualifications of robotic arc welding
   c. arc welding related disciplines
   d. robotic arc welding operators and technicians

24. Today, automation can mean solutions from a single robot to a full production line. (True or False)

25. It has been determined that only about _____ percent of the small to medium sized manufacturing companies have installed robots.

CHAPTER 28
Joint Design, Testing, and Inspection

Please answer the following questions by choosing the letter of the correct answer, circling true or false, or filling in the blanks.

1. Open roots are spaces between the edges of the members to be welded. (True or False)

2. Open roots in joints ensure complete penetration in welding the joint. (True or False)

3. _____ is the depth to which the base metal is melted and fused with the metal of the filler rod.

4. Whether a joint should be set up as an open or closed root depends upon all but which of the following?
   a. the thickness of the base metal   **b.** the kind of joint
   c. the nature of the job   **d.** the polarity of the electrode
   e. the position of welding

5. Edge joints are both strong and economical. (True or False)

6. Welding from both sides on a square groove butt joint materially increases the strength of a joint. (True or False)

7. Single V-groove butt joints, are superior to square-groove butt joints and are used a great deal for important work. (True or False)

8. Single-V butt joints provide for 100% penetration. (True or False)

9. Root penetration is not necessary on joints that are beveled on both sides and have a root opening. (True or False)

10. Match the types of corner joints to their best descriptions.
    1. Flush   **a.** Strong, butt fitup may be difficult   **a.** _____
    2. Half-open   **b.** Best used on light-gauge sheet metal   **b.** _____
    3. Full-open   **c.** Forms groove and permits root penetration   **c.** _____

11. Generally, product soundness and service life depend upon proper joint _____ and flawless welding.

12. All but which one of the following are parts of a ¾″ butt joint-groove weld.
    a. throat   **b.** heel   **c.** shoulder   **d.** overlap   **e.** face

13. Match the types of welds to their best descriptions.

1. Strength	**a.** Extends the entire length of a joint	**a.** _____
2. Seal Weld	**b.** Makes riveted joints leakproof	**b.** _____
3. Composite	**c.** Holds job parts together in preparation for welding	**c.** _____
4. Continuous	**d.** Carries the structural load	**d.** _____
5. Intermittent	**e.** Placed at intervals to reduce cost of noncritical work	**e.** _____
6. Tack	**f.** Meets load requirements and is also leakproof	**f.** _____

14. Match the welding positions to their descriptions.

1. Flat	**a.** Weld travel may be up or down	**a.** _____
2. Horizontal	**b.** Filler is deposited from upper side of joint; weld face is horizontal	**b.** _____
3. Vertical	**c.** Filler metal is deposited from underside of joint; weld face is horizontal	**c.** _____
4. Overhead	**d.** Filler is deposited on upper side of joint. The weld face is on top, with gravity assist.	**d.** _____

15. Metal extending above the surfaces of the plate is called _____.

16. High weld reinforcement increases the strength of a weld. (True or False)

17. It is usually the practice of a particular industry to develop a code. (True or False)

18. The distance from the root to the toe of a fillet weld is called the _____.

19. Throat thickness is the distance from the root to the _____ of the fillet weld.

20. Visual inspection is an adequate means of determining the quality of a weld. (True or False)

21. Different industries have different welding codes. (True or False)

22. A welder must be thoroughly informed about the demands of existing welding codes. (True or False)

23. The welding test that determines the correctness of the method of welding a specific project is the _____ qualification test.

24. Tests that judge the welder rather than the work are _____ qualification tests.

25. Evaluating some weld tests requires that the sample weld be destroyed beyond use. (True or False)

Name: _____ Date: _____

26. The welding test most commonly used by the welder is _____.

  a. nondestructive **b.** mechanical

  c. visual inspection **d.** destructive

27. Magnetic particle testing can be used with all metals. (True or False)

28. The trade name by which magnetic particle testing is often called is _____.

29. Magnetic particle testing detects the presence of surface and internal cracks as much as _____ of an inch below the weld surface.

30. Oil, water, and slag from the welding operation do not interfere with magnetic particle testing. (True or False)

31. Radiographic inspection makes use of _____ rays and X-rays.

32. Radiographic inspection methods are used to show the presence and type of macroscopic defects in the weld interior. (True or False)

33. Photographs for radiographic testing may be taken several feet away from work pieces that are hard to get close to. (True or False)

34. Penetrant inspection detects only _____ defects.

35. The fine white powder that soaks up the penetrant remaining on the workpiece to mark defects clearly is called a _____.

36. A penetrant test may also be conducted using a _____ material under special light instead of dyes and developers.

37. A closely controlled, rapid testing method that can probe deeply without damaging the weld is _____ inspection.

38. Magnetic particle testing is limited to magnetic materials such as steels and cast iron. (True or False)

39. Magnetic particle testing is one of the most easily used nondestructive tests. (True or False)

40. Eddy current testing may be used on both ferrous and nonferrous metals. (True or False)

41. Leak tests apply pneumatic or hydraulic _____ to determine a weldment's resistance to leakage.

42. Water can be used to detect leaks of all sizes. (True or False)

43. Brinell and Rockwell are common nondestructive _____ tests.

44. The test commonly used to determine the tensile strength of welds is
_____.

    a. free-bend   **b.** transverse shear

    c. nick-break   **d.** reducedsection tension

45. Tests for weld soundness include all of the following except which one?

    a. longitudinal shear   **b.** root-bend

    c. nick-break   **d.** face-bend

46. It is the general practice for code welding to qualify welders on the groove weld test in the 3G and 4G positions. (True or False)

47. In making a groove weld for test purposes, it is important to keep the tensile strength of the plate and the weld metal about the same. (True or False)

48. The most important part of a fillet test weld is the _____ pass.

49. All grinding and machining of test specimens must always be lengthwise of the specimen. (True or False)

50. A smooth finish of the test surface improves the chances of passing the test. (True or False)

51. It is not necessary to remove bead reinforcement from the test weld surface. (True or False)

52. No cracks or open defects of any kind are permitted on the surface of a test weld after it has been bent. (True or False)

53. Longitudinal and transverse shear tests determine the shearing strength of
_____ welds.

54. Etching reveals the penetration and soundness of a weld cross section. (True or False)

55. The Izod and Charpy tests are _____ tests.

56. _____ testing determines how well a weld can resist repetitive stress.

57. Match the principal weld defects to their best descriptions.

1. Cracking	**a.** Failure of filler and base metal to fuse and flow through to at the root of a joint	**a.** _____
2. Porosity	**b.** Burning away of base metal at weld toe	**b.** _____
3. Brittle welds	**c.** Presence of inclusions containing gases rather than solids	**c.** _____
4. Incomplete fusion	**d.** Failure of layers of weld metal to fuse at any point in the weld groove	**d.** _____
5. Dimensional defects	**e.** Contraction, warping, angular defects	**e.** _____
6. Incomplete penetration	**f.** Poor elongation, low-yield point, poor ductility, little resistance to stress	**f.** _____
7. Inclusions	**g.** Presence of linear ruptures of metal under stress	**g.** _____
8. Undercutting	**h.** Elongated or globular pockets of solid compounds	**h.** _____

58. _____ testing determines the tightness of welds in fabricated vessels.

59. Destructive testing gives an absolute measure of the strength of the sample tested. (True or False)

60. _____ testing is a good means of determining a welder's ability.

CHAPTER 29

Reading Shop Drawings

Please answer the following questions by choosing the letter of the correct answer, circling true or false, or filling in the blanks.

1. Welders do not rely on drawings. It is the engineers' responsibilities. (True or False)

2. Assembly drawings represent how all the parts come together to create the final assembly of the product. (True or False)

3. The term blueprint is replaced by the more modern term print. (True or False)

4. Match the types of drawing lines to their best descriptions.

   1. Extension   **a.** Represent edges or surfaces that cannot be seen in the finished product   **a.** _____

   2. Center   **b.** Point to a particular surface to show a dimension or a note   **b.** _____

   3. Hidden   **c.** Extend away from an object to indicate its size   **c.** _____

   4. Cutting plane   **d.** Show relationship of surfaces in one view with the same surfaces in other views   **d.** _____

   5. Leaders   **e.** Indicate a break in an object or that only part of an object is shown   **e.** _____

   6. Break   **f.** Sectional line to indicate where an imaginary cut is made   **f.** _____

   7. Projection   **g.** Dot-and-dash lines indicate the center of a circle or a part of a circle   **g.** _____

5. Match the following terms to the appropriate drawing lines in Fig. 29-1 on the next page.

   1. Phantom   **a.** _____
   2. Cutting plane   **b.** _____
   3. Dimension   **c.** _____
   4. Border   **d.** _____
   5. Leader   **e.** _____
   6. Object   **f.** _____
   7. Invisible edge   **g.** _____
   8. Extension   **h.** _____
   9. Center   **i.** _____

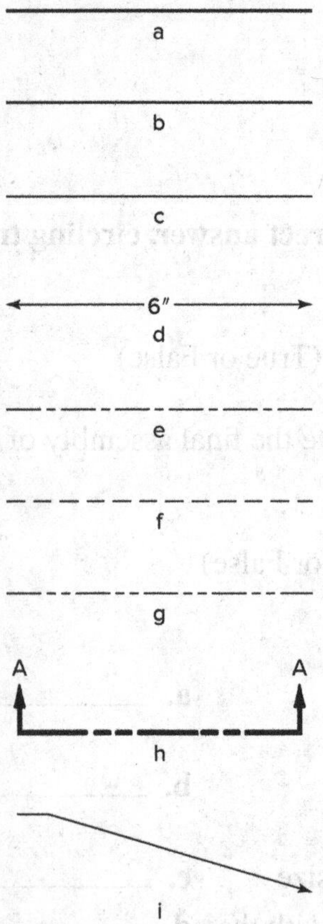

Fig. 29-1.

6. The metric measure of length is _____.
   a. liter   **b.** hectare   **c.** gram   **d.** meter

7. A metric measure of area is _____.
   a. liter   **b.** hectare   **c.** gram   **d.** meter

8. Volume may be measured in all but which of the following metric terms?
   a. cubic metres   **b.** liters   **c.** centimeters   **d.** milliliters

9. Mass may be measured in all but which of the following metric terms?
   a. grams   **b.** tones   **c.** square kilometers   **d.** kilograms

10. In the answer column, write the metric terms indicated by the following symbols.
   a. kg   **a.** _____
   b. ha   **b.** _____
   c. m   **c.** _____
   d. m^3 **d.** _____
   e. cm   **e.** _____
   f. cm^2 **f.** _____
   g. t   **g.** _____
   h. L   **h.** _____

11. In the answer blank, write 20 degrees, 15 minutes, and 8 seconds using the symbols for angular dimensions. _____

12. There are _____ degrees in a complete circle.

13. The permissible range of variation in the dimensions of a completed job is called the limits of

_____.

14. Match the figures from the drawing of the lock keeper in Fig. 29-2 to the appropriate measurements.

    1. Overall length   **a.** _____

    2. Thickness (in inches) of sheet brass   **b.** _____

    3. Recommended gauge thickness   **c.** _____

    4. Diameter of drilled holes   **d.** _____

    5. Number of drilled holes required   **e.** _____

    6. Distance from the center of the bottom drilled hole to the   **f.** _____
        bottom edge of the plate

    7. Distance from the center of the drilled holes to the left side of the keeper   **g.** _____

    8. Length of the rectangular hole   **h.** _____

    9. Width of the rectangular hole   **i.** _____

  10. Distance from the bottom of the rectangular hole to the bottom   **j.** _____
        edge of the keeper

  11. Distance from the top of the rectangular hole to the top edge of   **k.** _____
        the keeper

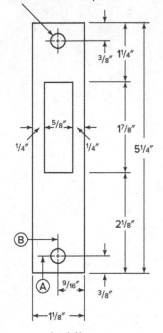

Drill ¼″ Hole 2 Required

Lock Keeper
Sheet Brass
No. 14 (0.064″) B. & S. GA.

Fig. 29-2.

CHAPTER 30
Welding Symbols

Please answer the following questions by choosing the letter of the correct answer, circling true or false, or filling in the blanks.*

1. The standard welding symbols have been developed by the _____.

2. Welding symbols give _____ welding information.

3. Weld symbols are elements of welding symbols. (True or False)

4. Specifications for welding are given in the _____ of the reference line.

5. Match the following types of welds to the corresponding symbols shown below.

   1. Slot   **a.** _____
   2. Edge   **b.** _____
   3. V groove   **c.** _____
   4. Fillet   **d.** _____
   5. Seam   **e.** _____
   6. Bevel groove   **f.** _____
   7. Square groove   **g.** _____

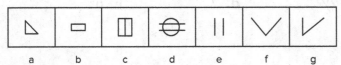

Fig. 30-1.

6. Match the following descriptions to the appropriate supplementary symbols shown below.

   1. Field weld   **a.** _____
   2. All around weld   **b.** _____
   3. Melt-through   **c.** _____

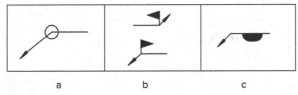

Fig. 30-2.

*Figs. 30-1 and 30-2 Adapted from A2.4-2007, figure 50. American Welding Society, 2007.
Figs. 30-3 to 30-5 and 30-11 from A2.4–2007, pages 114–115. Copyright © 2007 American Welding Society. Used with permission from the American Welding Society (AWS), Miami, Florida.

7. The type of weld indicated is _____.

   a. fillet   **b.** double fillet   **c.** seam   **d.** plug   **e.** double slot

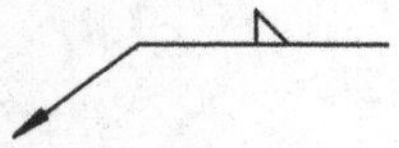

Fig. 30-3.

8. The type of groove indicated is _____.

   a. U   **b.** J   **c.** double-U   **d.** double-J   **e.** square

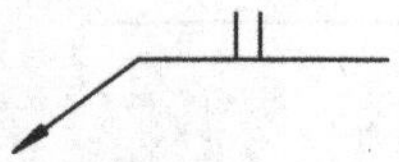

Fig. 30-4.

9. The welding should be done arrow side. (True or False)

Fig. 30-5.

10. The welding on the _____ side are tack welds.

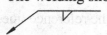

Fig. 30-6.

11. The distance from center to center of the tack welds is called the _____.

12. The depth of the groove, arrow-side weld is _____.

   a ½″   **b.** ³⁄₃₂″   **c.** 2½″   **d.** 2″   **e.** 6″

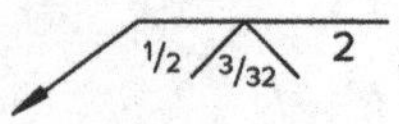

Fig. 30-7.

13. The size of the fillet weld, other-side weld is _____.

   a. 3″   **b.** 5″   **c.** ⅜″   **d.** ¼″   **e.** 2″

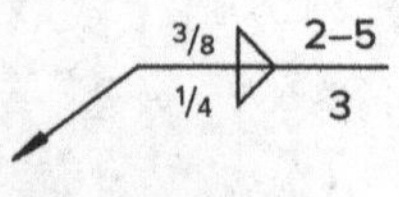

Fig. 30-8.

14. The size of the fillet weld, arrow-side weld is _____.

   a. 3″   **b.** 5″   **c.** ⅜″   **d.** ¼″   **e.** 2″

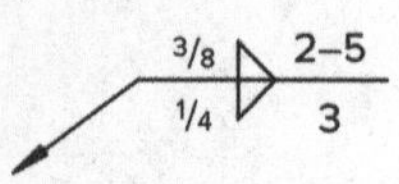

Fig. 30-9.

15. Multiple reference lines should not be used when a particular order in welding sequence is needed. (True or False)

16. The groove angle of the groove weld is _____ degrees.

Fig. 30-10.

17. The black flag indicates a _____ weld.

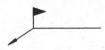

Fig. 30-11.

18. The desired shape of the weld is shown on the welding symbol by use of a _____ symbol.

CHAPTER 31

Welding and Bonding of Plastics

Please answer the following questions by choosing the letter of the correct answer, circling true or false, or filling in the blanks.

1. The synthetic polymers (polymeric materials) have been used in the world economy for over _____ years.

2. Plastics are useful because of its excellent properties—corrosion resistant, light weight, and fatigue resistant. (True or False)

3. Colors should always be used to identify a type of plastic. (True or False)

4. Plastics have properties similar to aluminum in that they have different specific _____ and surface hardnesses.

5. Thermoplastics soften and/or melt when heated and can be welded. (True or False)

6. There are two basic types of plastics: (1) thermosetting plastics and (2) thermoplastics. (True or False)

7. _____ plastics are transparent.

8. For welding thermoplastics, which of the following is not correct?
   a. similar to gas-welding metals
   b. fast, but somewhat expensive
   c. permanent
   d. possible in all positions

9. All layout work on thermoplastics should be done with all but which of the following?
   a. lead pencil
   b. soapstone
   c. a china marker
   d. a scribe

10. Thermoplastics should be preshrunk. (True or False)

11. Cutting thermoplastics with a _____ saw produces less heat than other cutting methods.

12. Light-gauge plastic sheets may be sheared without producing stress marks if they are cooled slightly. (True or False)

13. As with metals, plastics are welded by applying enough localized heat to produce _____ of the areas that are to be joined.

14. Methods of welding plastics include all but which of the following?
    a. hot-gas
    b. infrared
    c. conduction
    d. microwave
    e. spin

15. Stretching may be caused by all but which one of the following?

    a. leaning the welding rod away from the direction of welding

    b. too little pressure on the welding rod

    c. plastic residue on the shoe

    d. insufficient cooling between multipass welds

16. Reduce distortion by speed welding with _____ welding rods.

17. Because PVC is a poor heat conductor, choose backup materials that conduct heat well. (True or False)

18. Environmental stress cracking results from _____ attack that would not normally be a threat.

19. All plastic welding torches are electrically heated. (True or False)

20. Plastic welding torches may provide a heat range up to _____ degrees F.

21. Compressed air or _____ gas is used as the hot gas medium.

22. When welding plastics with electric torches, turn the gas on first and off last. (True or False)

23. When welding plastics with an electric torch, _____ the gas volume to increase the welding temperature.

24. When welding with a gas-heated, plastics welding torch, reduce the volume of the welding gas or _____ the pressure of the heating gas to raise the welding temperature.

25. At the end of the welding operation with the gas-heated, plastics welding torch, always turn off the heating flame before shutting off the welding gas. (True or False)

26. Plastic welds that are equal to less than _____% of the base material strength are considered unsatisfactory.

27. Welded joints are the sites of potential weakness in a plastic structure. (True or False)

28. Match the following defects of plastic welds to their possible causes.

1. Porosity	**a.** Too little root gap	**a.** _____
2. Poor penetration	**b.** Shrinking of base material	**b.** _____
3. Scorching	**c.** Stretching the welding rod	**c.** _____
4. Warping	**d.** Rod and base material of different composition	**d.** _____
5. Stress cracking	**e.** Base material too cold	**e.** _____

29. Match the following plastic weld defects to the appropriate corrective steps.

1. Scorching	a. Weld rapidly.	a. _____
2. Distortion	b. Use small rod at root, large rods at top of weld.	b. _____
3. Warping	c. Back up weld with metal.	c. _____
4. Poor appearance	d. Increase air flow.	d. _____
5. Poor fusion	e. Use proper rod angle.	e. _____

30. Destructive tests for plastic welds include all but which one of the following?

    a. tensile   **b.** bending   **c.** chemical   **d.** impact

31. The most effective way to test welds on plastic pipe is to subject the fabrication to the _____ test.

32. To detect pores and cracks not visible in plastic welds by any other type of inspection, use _____.

    a. tensile tests   **b.** impact tests   **c.** spark coil tests   **d.** radiography

33. When welding plastics you must do all but which one of the following?

    a. place large beads at the base of the weld

    b. be sure the rod remains round

    c. be careful not to char the rod or base material

    d. do not use flammable gases

34. The most efficient, but expensive, method of inspecting the internal characteristics of a plastic weld is _____.

35. For welding plastics, all but which of the following is true about the filler rod?

    a. should be the same length as the weld

    b. should be forced ahead of the weld to conserve materials

    c. must not be stretched

    d. must be of the same composition as the base material

36. A tensile strength value of _____ percent of the base material is considered acceptable.

37. Heat for welding plastics is supplied by a low-heat torch flame. (True or False)

38. Heating gases for plastic welding may be all but which one of the following?

    a. compressed air   **b.** nitrogen

    c. oxygen   **d.** inert gas

39. Decreasing the volume of welding gas for joining plastics _____ the temperature of the gas.

40. Torch heating capacity may be altered by changing the heating element and _____.

41. For tack welding plastic joints, all but one of the following is true. Which one is incorrect?

    a. assembles parts quickly

    b. eliminates the need for jigs and fixtures

    c. contributes to good fusion

    d. ensures proper part alignment

42. Plastics do not need to be clean and dry prior to welding and during the welding operation. (True or False)

43. Successful beading on plastics depends upon the proper combination of _____ and heat.

44. A slight yellowing of the filler rod and base material is caused by _____.

45. One of the most critical parts of making a quality weld is maintaining the proper temperature. (True or False)

46. For high speed welding of plastics, all but one of the following is true. Which one is incorrect?

    a. is even faster with a 45-degree torch angle

    b. does not produce flowlines

    c. forms a higher crown than hand welding

    d. must keep moving to avoid rod softening

47. When joining plastic sheet, the welding proceeds only in a _____ direction.

48. _____ welding of plastic pipe produces a joint that is stronger than both the pipe and the fitting.

49. For plastic pipe all but which one of the following are true?

    a. must never be stored in the sun   **b.** must be carefully cut

    c. may be heated and bent   **d.** may be installed in a cinder fill

50. Adhesive dispersion is commonly used in plastic joining. (True or False)

CHAPTER 32
Safety

Please answer the following questions by choosing the letter of the correct answer, circling true or false, or filling in the blanks.

1. A.c. transformer welding machines may be attached to a.c. lighting circuits. (True or False)

2. The work lead to the work is adequate grounding for the welding machine. (True or False)

3. A welding machine may be moved safely while it is connected to the power supply. (True or False)

4. When several lengths of cable are coupled together, connectors on both work and electrode lines should be insulated. (True or False)

5. Size 2 copper cable may be safely used with 200 amperes of welding current with a 60% duty cycle welding machine. (True or False)

6. A container of water should be available for cooling hot electrode holders. (True or False)

7. The power supply must be turned off for changing TIG electrodes or threading MIG/MAG equipment. (True or False)

8. Industrial alcohol cleans commutators quickly and safely. (True or False)

9. After welding, the entire welding machine may be blown out with clean, dry, compressed air. (True or False)

10. Air filters may cause overheating. (True or False)

11. The infrared and ultraviolet rays produced by the electric arc may "sunburn" an unprotected welder. (True or False)

12. Anyone over four feet away from the electric arc is safe from any harmful effects of its rays. (True or False)

13. Before welding, welders should use a systematic approach to make sure they have not forgotten any of their PPE. (True or False)

14. Helmets and goggles should be sterilized before being transferred from one operator to another. (True or False)

15. For better vision during weld cleaning, helmets and goggles maybe raised if the power-driven brush has a hood guard. (True or False)

16. The simplest form of ear protection is the large device that fits directly in your ear canal. (True or False)

17. One of the most effective ways to protect you from the hazardous effects of welding fumes is to keep your head out of the fume plume. (True or False)

18. High temperature and humidity increase the hazards of electric shock. (True or False)

19. Cotton clothing ignites more easily than wool. (True or False)

20. Even the ears must be protected during overhead welding. (True or False)

21. The low voltages required for arc welding eliminate the danger of severe electric shock. (True or False)

22. Keep electrical cable splices or insulation repairs within 10 feet of the electrode holder. (True or False)

23. The mixture of flammable gases and oxygen in oxyfuel welding torches is highly explosive. (True or False)

24. Gas cylinders may be lifted with a crane only when secured to a cradle or platform. (True or False)

25. Empty cylinders require no closing or capping. (True or False)

26. Well-constructed cylinders may be stored outside in an open area. (True or False)

27. Store acetylene cylinders on their sides to keep acetone and packing material from settling to the bottom. (True or False)

28. A regulating device must be installed between an acetylene cylinder and the torch. (True or False)

29. Crack acetylene cylinders before attaching regulators. (True or False)

30. Use a match to test for acetylene leak. (True or False)

31. Oxygen does not burn. (True or False)

32. Oxygen may be substituted for compressed air. (True or False)

33. Valves on oxygen cylinders must remain partly closed for welding. (True or False)

34. An oxygen hose may be attached to any torch hose connection. (True or False)

35. For changing torches, gases must be shut off at the pressure regulators. (True or False)

36. When oxyacetylene welding is finished, the flame is extinguished by closing both cylinder valves at the same time. (True or False)

37. When welding is stopped for only a few minutes, it is all right to close only the torch valves. (True or False)

38. If torch valves freeze or become hard to operate, remove and oil the valve stem. (True or False)

39 Clean clogged nozzle holes with any sharp drill. (True or False)

40. It is permissible to use one hose for several different gases, as long as it is well cleaned between uses. (True or False)

41. Hose splices should be securely taped. (True or False)

42. Thorough visual inspection should determine whether or not a section of hose is safe for use after a flashback. (True or False)

43. A standard pressure regulator may be used for several different gases. (True or False)

44. The adjusting screw on a regulator must be released before the gas is turned on. (True or False)

45. A fire extinguisher should be part of a welder's standard equipment. (True or False)

46. When welding inside a confined space, such as a boiler, set both cylinders inside so they may be turned off quickly. (True or False)

47. Concrete is a safe working surface. (True or False)

48. Too much oxygen pressure can blow sparks out of the work area. (True or False)

49. Touching the torch tip to the work may cause preignition and backfire. (True or False)

50. Even when a torch is improperly handled, flashback rarely occurs. (True or False)

51. Oxygen from a welding cylinder may also be used to aid in ventilation. (True or False)

52. Wooden floors should be covered with noncombustible materials or dampened before welding is begun. (True or False)

Appendix A

Job Outlines

Table 7-2. Job Outline: OFC Practice   136

Table 8-4. Job Outline: Gas Welding Practice: Jobs 8-J1–J38   137

Table 9-3. Job Outline: Advanced Gas Welding and Braze Welding Practice: Jobs 9-J39–J49   138

Table 10-6. Job Outline: Soldering and Brazing Practice: Jobs 10-J50 and J51   138

Table 13-6. Job Outline: Shielded Metal Arc Welding Practice: Jobs 13-J1–28 (Plate)   139

Table 14-1. Job Outline: Shielded Metal Arc Welding Practice: Jobs 14-J26–J42 (Plate)   140

Table 15-5. Job Outline: Shielded Metal Arc Welding Practice: Jobs 15-J43–J55 (Plate)   141

Table 16-11. Job Outline: Shielded Metal Arc Welding Practice (Pipe)   142

Table 17-11. Job Outline: Air Carbon Arc Cutting: Jobs 17-J6–J7   144

Table 19-5. Job Outline: Gas Tungsten Arc Welding Practice (Plate)   145

Table 20-2. Job Outline: Gas Tungsten Arc Welding Practice (Pipe)   146

Table 22-8. Job Outline: Gas Metal Arc Welding Practice with Solid Core Wire (Plate)   148

Table 23-10. Job Outline: Flux Cored Arc Welding Practice with Gas Shielded Electrodes (Plate)   150

Table 23-11. Job Outline: Flux Cored Arc Welding Practice with Self-Shielded Electrodes   151

Table 23-12. Job Outline: Semiautomatic Application of the Submerged Arc Welding Process   152

Table 24-2. Job Outline: Gas Metal Arc Welding Practice with Solid Wire (Pipe)   153

Table 7-2 Job Outline: OFC Practice

Job Number	Type of Job	Material[1]		Gas Pressure (p.s.i.)[2]		Diameter of Cutting Orifice (in.)	Text Reference	AWS SENSE Reference[3]
		Type	Size (in.)	Acet.	Oxy.			
7-J0	Setting up equipment and closing down equipment	OFC system	NA	5	20	0.030–0.060	213	Entry level
7-J1	Straight line and bevel cutting	CS	¼ × 7 × 8	5	20	0.030–0.060	218	Entry level
7-J2	Laying out and cutting odd shapes	CS	¼ × 8-½ × 11	5	20	0.030–0.060	222	Entry level
7-J3	Cutting cast iron	Gray CI	Available	5	25–60	0.030–0.060	224	Entry level

[1]CS = carbon steel, CI = cast iron.

[2]For specifics on your equipment, check your tip size and use the manufacturer's recommended pressure settings.

[3]Entry-level weld removal (weld washing) machine cutting, and advanced and expert level pipe cutting are not covered in this chapter.

Table 8-4 Job Outline: Gas Welding Practice: Jobs 8-J1–J38

Job Order[1]	Text No.	Joint	Operation	Material Type	Thickness	Filler Rod Type	Size	Welding Position[2]	Text Reference
1st	8-J1	Flat plate	Autogenous welding	Mild steel	1/8	None		1	213
2nd	8-J2	Flat plate	Autogenous welding	Mild steel	1/8	None		3	214
3rd	8-J3	Flat plate	Beading	Mild steel	1/8	Mild steel RG45	3/32	1	214
4th	8-J9	Edge joint	Flange welding	Mild steel	1/8	Mild steel RG45	3/32	1	216
5th	8-J10	Corner joint	Groove welding	Mild steel	1/8	Mild steel RG45	3/32	1	217
6th	8-J4	Flat plate	Beading	Mild steel	1/8	Mild steel RG45	3/32	3	218
7th	8-J11	Edge joint	Flange welding	Mild steel	1/8	Mild steel RG45	3/32	3	218
8th	8-J12	Corner joint	Groove welding	Mild steel	1/8	Mild steel RG45	3/32	3	219
9th	8-J5	Flat plate	Autogenous welding	Mild steel	1/8	None		2	219
10th	8-J6	Flat plate	Beading	Mild steel	1/8	Mild steel RG45	3/32	2	219
11th	8-J13	Edge joint	Flange welding	Mild steel	1/8	Mild steel RG45	3/32	2	219
12th	8-J14	Corner joint	Groove welding	Mild steel	1/8	Mild steel RG45	3/32	2	219
13th	8-J7	Flat plate	Autogenous welding	Mild steel	1/8	None	3/32	4	219
14th	8-J8	Flat plate	Beading	Mild steel	1/8	Mild steel RG45	3/32	4	220
15th	8-J15	Edge joint	Flange welding	Mild steel	1/8	Mild steel RG45	3/32	4	220
16th	8-J16	Corner joint	Groove welding	Mild steel	1/8	Mild steel RG45	3/32	4	220
17th	8-J19	Lap joint	Fillet welding	Mild steel	1/8	Mild steel RG45	3/32	1	220
18th	8-J20	Lap joint	Fillet welding	Mild steel	1/8	Mild steel RG45	3/32	3	220
19th	8-J17	Square butt joint	Groove welding	Mild steel	1/8	Mild steel RG45	3/32	1	222
20th	8-J18	Square butt joint	Groove welding	Mild steel	1/8	Mild steel RG45	3/32	3	222
21st	8-J21	Lap joint	Fillet welding	Mild steel	1/8	Mild steel RG45	3/32	2	222
22nd	8-J23	T-joint	Fillet welding	Mild steel	1/8	Mild steel RG45	3/32	2	222
23rd	8-J22	Lap joint	Fillet welding	Mild steel	1/8	Mild steel RG45	3/32	4	222
24th	8-J24	T-joint	Fillet welding	Mild steel	1/8	Mild steel RG45	3/32	3	222
25th	8-J25	T-joint	Fillet welding	Mild steel	1/8	Mild steel RG45	3/32	4	222
26th	8-J26	Flat plate	Beading—backhand	Mild steel	3/16	Steel, high test RG-60	1/8	1	223
27th	8-J27	Flat plate	Beading	Mild steel	3/16	Steel, high test RG-60	1/8	3	223
28th	8-J28	Lap joint	Fillet welding—backhand	Mild steel	3/16	Steel, high test RG-60	1/8	1	223
29th	8-J29	Flat plate	Beading—backhand	Mild steel	3/16	Steel, high test RG-60	1/8	2	223
30th	8-J30	Lap joint	Fillet welding—backhand	Mild steel	3/16	Steel, high test RG-60	1/8	2	223
31st	8-J31	Lap joint	Fillet welding	Mild steel	3/16	Steel, high test RG-60	1/8	3	223
32nd	8-J32	Beveled butt joint	Groove welding—backhand	Mild steel	3/16	Steel, high test RG-60	1/8	1	223
33rd	8-J33	Beveled butt joint	Groove welding	Mild steel	3/16	Steel, high test RG-60	1/8	3	224
34th	8-J34	Beveled butt joint	Groove welding	2- to 4-in. pipe	Standard	Steel, high test RG-60	1/8 or 5/32	IG-roll	226
35th	8-J35	Beveled butt joint	Groove welding—forehand	2- to 4-in. pipe	Standard	Steel, high test RG-60	1/8 or 5/32	5G	226
36th	8-J37	Beveled butt joint	Groove welding—backhand	2- to 4-in. pipe	Standard	Steel, high test RG-60	5/32 or 3/16	3G	226
37th	8-J36	Beveled butt joint	Groove welding—backhand	Mild steel plate	3/16	Steel, high test RG-60	1/8 or 5/32	5G	228
38th	8-J38	Beveled butt joint	Groove welding—backhand	2- to 4-in. pipe	Standard	Steel, high test RG-60	1/8 or 5/32	2G	229

[1]It is recommended that the student do the jobs in this order. In the text, the jobs are grouped according to the type of operation to avoid repetition.

[2]1 = flat, 2 = horizontal, 3 = vertical, 4 = overhead, 5 = multiple positions pipe.

Note: The AWS S.E.N.S.E. does not cover gas welding practice.

Job No.	Joint	Operation	Material Type	Thickness	Filler Rod Type	Size	Welding Position[1]	Text Reference
9-J39	Flat plate	Beading (braze welding)	Mild steel	⅛	Bronze RCuZn-C	⅛	1	241
9-J40	Lap joint	Fillet (braze welding)	Mild steel	⅛	Bronze RCuZn-C	⅛	1	241
9-J41	T-joint	Fillet (braze welding)	Mild steel	⅛	Bronze RCuZn-C	⅛	2	241
9-J42	Beveled-butt joint	Groove (braze welding)	Mild steel	3⁄16	Bronze RCuZn-C	⅛	1	241
9-J43	Flat casting	Beading (braze welding)	Cast iron	3⁄16	Bronze RCuZn-C	⅛	1	241
9-J44	Beveled-butt joint	Groove (braze welding)	Cast iron	3⁄16 to ¼	Bronze RCuZn-C	⅛	1	241
9-J45	Casting	Beading (fusion)	Cast iron	3⁄16 to ¼	Cast iron RCI	3⁄16 or ¼	1	245
9-J46	Beveled-butt joint	Groove welding (fusion)	Cast iron	¼ to ½	Cast iron RCI	3⁄16 or ¼	1	245
9-J47	Flat plate	Beading	Sheet aluminum	⅛	Aluminum R1100 or R4043	3⁄32 or ⅛	1	248
9-J48	Outside corner joint	Groove welding	Sheet aluminum	⅛	Aluminum R1100 or R4043	⅛	1	248
9-J49	Square butt joint	Groove welding	Sheet aluminum	⅛	Aluminum R1100 or R4043	⅛	1	248

Note: It is recommended that students complete the jobs in the order shown. At this point continue welding practice with those other metals described in the chapter that are available in the school shop. Hard facing should also be practiced.

It is very important that you become proficient in gas cutting. (See Chapter 7.) Practice should include straight line, shape, and bevel cutting with both the hand cutting torch and the machine cutting torch. This is a skill that is also necessary for the electric arc welder.

The AWS S.E.N.S.E. does not cover braze welding or advanced gas welding practice.

[1]1 = flat, 2 = horizontal, 3 = vertical, 4 = overhead, 5 = multiple positions pipe.

Job No.	Joint	Operation TB	Material Type	Diam. (in.)	Filler Rod/Flux Type	Size	Welding Position[1]	Text Reference
10-J50	Pipe fitting, socket	Solder copper tubing	Copper	½ to 1	50–50 or 95–5, appropriate flux	Available	2, 5, and 6	265
10-J51	Pipe fitting, socket	Braze copper pipe and tubing	Copper	½ to 1½	BCuP or BAg, appropriate flux	Available	2, 5, and 6	277

Note: It is recommended that students complete the jobs in the order shown.
*The AWS S.E.N.S.E. program does not cover soldering or brazing.
[1]2 = horizontal, 5 and 6 = multipositions.

Table 13-6 Job Outline: Shielded Metal Arc Welding Practice: Jobs 13-J1–28 (Plate)

Job No.	Joint	Type of Weld	Position	Type of Electrode DCEP	DCEN	A.C.	AWS Specif. No. d.c.	AWS and/or	AWS Specif. No. a.c.	Text Reference	AWS S.E.N.S.E. Level
13-J1	Flat plate	Striking the arc—short beading	Flat 1		X		E6013		E6013	349	NA
13-J2	Flat plate	Beading—stringer	Flat 1		X	X	E6013		E6013	351	NA
13-J3	Flat plate	Beading—weaved	Flat 1		X	X	E6013		E6013	354	NA
13-J4	Flat plate	Beading—stringer	Flat 1	X		X	E6010		E6011	356	Entry Level
13-J5	Flat plate	Beading—weaved	Flat 1	X		X	E6010		E6011	358	Entry Level
13-J6	Edge joint	Edge	Flat 1		X	X	E6013		E6013	359	NA
13-J7	Edge joint	Edge	Flat 1	X		X	E6010		E6011	361	Entry Level
13-J8	Lap joint (plates flat)	Fillet—single pass	Hor. 2F		X	X	E6013		E6013	363	NA
13-J9	Lap joint (plates flat)	Fillet—single pass	Hor. 2F	X		X	E6010		E6011	365	Entry Level
13-J10	Flat plate	Beading—stringer	Hor. 2	X		X	E6010		E6011	366	Entry Level
13-J11	Flat plate	Beading—stringer—travel down	Ver. 3	X		X	E6010		E6011	368	NA
13-J12	Lap joint	Fillet—single pass—travel down	Ver. 3F		X	X	E6013		E6013	370	NA
13-J13	Lap joint (plates vert.)	Fillet—single pass	Hor. 2F		X	X	E6013		E6013	372	NA
13-J14	T-joint	Fillet—single pass	Flat 1F		X	X	E6013		E6013	374	NA
13-J15	T-joint	Fillet—weaved multipass	Flat 1F		X	X	E6013		E6013	376	NA
13-J16	T-joint	Fillet—single pass	Flat 1F	X		X	E6010		E6011	378	Entry Level
13-J17	T-joint	Fillet—weaved multipass	Flat 1F	X		X	E6010		E6011	380	Entry Level
13-J18	Flat plate	Beading-stringer—travel up	Ver. 3	X		X	E6010		E6011	381	Entry Level
13-J19	Flat plate	Beading-weaved—travel up	Ver. 3	X		X	E6010		E6011	383	Entry Level
13-J20	Flat plate	Beading-weaved—travel up	Ver. 3		X	X	E6013		E6013	385	NA
13-J21	Lap joint	Fillet—single pass-travel up	Ver. 3F	X		X	E6010		E6011	386	Entry Level
13-J22	T-joint	Fillet—single pass	Hor. 2F		X	X	E6013		E6013	389	NA
13-J23	T-joint	Fillet—stringer—multipass and weave pass	Hor. 2F		X	X	E6013		E6013	391	NA
13-J24	T-joint	Fillet—single pass	Hor. 2F	X		X	E6010		E6011	393	Entry Level
13-J25	T-joint	Fillet—stringer—multipass and weave pass	Hor. 2F	X		X	E6010		E6011	394	Entry Level

Table 14-1 Job Outline: Shielded Metal Arc Welding Practice: Jobs 14-J26–J42 (Plate)

Job No.	Joint	Type of Weld	Position	Type of Electrode DCEP	DCEN	a.c.	AWS Specif. No. d.c.	and/or	AWS Specif. No. a.c.	Text Reference	AWS S.E.N.S.E. Reference
14-J26	Flat plate	Beading—stringer	Over. (4)	X		X	E6010		E6011	400	NA
14-J27	Flat plate	Beading—weaved	Over. (4)	X		X	E6010		E6011	402	NA
14-J28	Single-V butt joint backing bar	Groove welding—weaved—multipass	Flat (1G)	X		X	E6010		E6011	404	Entry level
14-J29	T-joint	Fillet welding—weaved—multipass	Hor. (2F)	X		X	E6010		E6011	406	Entry level
14-J30	Single-V butt joint backing bar	Groove welding—stringer—multipass	Hor. (2G)	X		X	E6010		E6011	408	Entry level
14-J31	Square butt joint	Groove welding—multipass	Flat (1G)	X		X	E6010		E6011	410	Entry level
14-J32	Outside corner joint	Groove welding—weaved—multipass	Flat (1G)	X		X	E6010		E6011	411	Entry level
14-J33	Single-V butt joint	Groove welding—weaved—multipass	Flat (1G)	X		X	E6010		E6011	414	Entry level
14-J34	T-joint	Fillet welding—stringer—multipass	Hor. (2F)	X		X	E7016		E7016	417	Entry level
14-J35	T-joint	Fillet welding—single pass—travel up	Ver. (3F)	X		X	E6010		E6011	419	Entry level
14-J36	T-joint	Fillet welding—weaved—multipass—travel up	Ver. (3F)	X		X	E6010		E6011	421	Entry level
14-J37	T-joint	Fillet welding—weaved—multipass—travel up	Ver. (3F)	X		X	E7018		E7018	423	Entry level
14-J38	Single-V butt joint backing bar	Groove welding—weaved—travel up	Ver. (3G)	X		X	E6010		E6011	425	Entry level
14-J39	Square butt joint	Groove welding—multipass—travel up	Ver. (3G)	X		X	E6010		E6011	427	Entry level
14-J40	Outside corner joint	Groove welding—weaved—multipass—travel up	Ver. (3G)	X		X	E6010		E6011	429	Entry level
14-J41	T-joint	Fillet welding—weaved—multipass	Hor. (2F)	X		X	E7018		E7018	431	Entry level
14-J42	Single-V butt joint backing bar	Groove welding—weaved—multipass—travel down	Ver. (3G)	X		X	E6010		E6011	433	Entry level

NA = not applicable.

Table 15-5 Job Outline: Shielded Metal Arc Welding Practice: Jobs 15-J43–J55 (Plate)

Job No.	Joint	Type of Weld	Position	DCEP	DCEN	a.c.	AWS Specif. No. d.c.	and/or	AWS Specif. No. a.c.	Text Reference	AWS S.E.N.S.E. Reference[1]
15-J43	Single-V butt joint	Groove welding—weaved—multipass—travel up	Ver. (3G)	X		X	E6010 E7016–18		E6011 E7016–18	440	NA
15-J44	T-joint	Fillet welding—single pass	Over. (4F)	X		X	E6010		E6011	443	Entry level
15-J45	T-joint	Fillet welding—stringer multipass	Over. (4F)	X		X	E6010		E6011	445	Entry level
15-J46	Lap joint	Fillet welding—single pass	Over. (4F)	X		X	E6010		E6011	446	Entry level
15-J47	Lap joint	Fillet welding—single pass	Over. (4F)	X		X	E7016, 7018		E7016, 7018	447	Entry level
15-J48	Single-V butt joint	Groove welding—weaved—multipass—travel down	Ver. (3G)	X		X	E6010		E6011	450	NA
15-J49	T-joint	Fillet welding—multipass	Over. (4F)	X		X	E6010		E6011	454	Entry level
15-J50	T-joint	Fillet welding—weaved—multipass	Over. (4F)	X		X	E6010		E6011	455	Entry level
15-J51	Single-V butt joint	Groove welding—stringer—multipass	Hor. (2G)	X		X	E6010 E7016–18		E6011 E7016–18	458	NA
15-J52	Fittings	Fillet welding—single pass	Hor. (2F)	X		X	E7016–18		E7016–18	458	Entry level
15-J53	Fittings	Fillet welding—single pass	Hor. (2F)	X		X	E6010		E6011	459	Entry level
15-J54	Single-V butt joint backing bar	Groove welding—stringer multipass	Over. (4G)	X		X	E6010 E7016–18		E6011 E7016–18	461	Entry level[2]
15-J55	Single-V butt joint	Groove welding—weaved multipass	Over. (4G)	X		X	E6010 E7016–18		E6011 E7016–18	463	NA

NA = not applicable.

[1] AWS S.E.N.S.E. does not cover open-root joints on plate.

[2] For entry level 2G–3G with backing, bend tested (open-root joints also qualify). 4G only visual inspection (no bend test required). To qualify for advanced level on plate, the student must also perform fillet and groove welds in all positions with stainless-steel electrodes on carbon steel plate or stainless-steel plate.

Table 16-11 Job Outline: Shielded Metal Arc Welding Practice (Pipe)

Job No.	Type of Joint	Type of Weld	Welding Position	Welding Technique	Pipe Specifications Dia. (in.)	Weight	Electrode Specifications Type[1]	Size (in.)	Polarity	Text References This Unit	Other Jobs	AWS S.E.N.S.E.
16-J1	Pipe	Surface	Horizontal roll (1)	Surface stringer & weaved.	4-10	Schedule 40	E6010 E7018	3/32, 1/8, & 5/32	EP	513	13-J4 & J5	NA
16-J2	Butt	V-Groove	Horizontal roll (1G)	1st pass stringer; 3 passes weaved.	6-10	Schedule 40	E6010 E7018	1P-1/8-6010 Others 3/32-6010 1/8-7018	EP	513	14-J33	NA
16-J3	Butt	V-Groove	Vertical fixed (2G)	7 passes stringer.	6-10	Schedule 40	E6010 E7018	1 to 3P-1/8-6010 4 to 7P-3/32-6010 2 to 7P-1/8-7018	EP	514	15-J51	Advanced level
16-J4	Butt	V-Groove	Vertical fixed (2G)	7 passes stringer, Cover pass weave.	6-10	Schedule 40	E6010 E7018	1 to 3P-1/8-6010 4 to 7P-3/32-6010 Weave 3/32-6010 2 to 7P-1/8-7018	EP	514	13-J24	Advanced level
16-J5	Butt	V-Groove	Horizontal fixed (5G)	1st pass stringer, 2 passes weaved; travel up.	6-10	Schedule 40	E6010 E7018	1P-1/8-6010 2P-3/32-6010 3P-3/32-6010 2 to 3P-7018	EP	515	15-J43 & J55	Advanced level
16-J6	Butt	V-Groove	Horizontal fixed (5G)	1st pass stringer, 3 passes weaved; travel down.	6-10	Schedule 40	E6010	1P-1/8 2P-3/32 3-4P-3/16	EP	518	15-J48 & J55	Advanced level
16-JQT1[2]	Butt with backing	V-Groove	45° from horizontal fixed (6G)	Stringer passes.	6-8 M-1/P-1 group 1 or 2	Schedule 80	E7018	3/32 or 1/8 all passes	EP	520	NA	Advanced level
16-JQT2[3]	Butt without backing	V-Groove	45° from horizontal fixed (6G)	Stringer passes.	6-8 M-1/P-1 group 1 or 2	Schedule 80	E6010 E7018	1P-1/8-6010 Subsequent passes 3/32-1/8-7018	EP	521	NA	Advanced level
16-JQT3[4]	Butt without backing	Bevel-Groove	45° from horizontal fixed with restricting (6GR)	Stringer passes.	12 M-1/P-1 group 1 or 2	Schedule 80	E6010 E7018	1P-1/8-6010 Subsequent passes 3/32-1/8-7018	EP	521	NA	NA
16-J7	90° branch	Groove fillet	Top (branch)	Multipass stringer.	4-10 Small to large. Size on size.	Schedule 40	E6010	1/8 & 3/32	EP	522	14-J33 13-J25	Advanced level
16-J8	90° branch	Groove fillet	Horizontal (branch)	Multipass: 1st pass stringer, others weaved.	4-10 Small to large. Size on size.	Schedule 40	E6010	1/8 & 3/32	EP	522	15-J43 14-J36	Advanced level
16-J9	90° branch	Groove fillet	Bottom (branch)	Multipass: Last pass weaved.	4-10 Small to large.	Schedule 40	E6010	1/8 & 3/32	EP	522	15-J55 13-J24	Advanced level

					Pipe Specifications		Electrode Specifications			Text References		AWS
Job No.	Type of Joint	Type of Weld	Welding Position	Welding Technique	Dia. (in.)	Weight	Type[1]	Size (in.)	Polarity	This Unit	Other Jobs	S.E.N.S.E.
16-J10	45° branch	Groove fillet	Top (branch)	Multipass stringer.	4–10 Small to large. Size on size.	Schedule 40	E6010	⅛ & 5/32	EP	522	14-J33 13-J25	Advanced level
16-J11	45° branch	Groove fillet	Horizontal (branch)	Multipass: 1st pass stringer; others weaved.	4–10 Small to large. Size on size.	Schedule 40 or 80	E6010	⅛ & 5/32	EP	522	15-J43 14-J36	Advanced level
16-J12	45° branch	Groove fillet	Bottom (branch)	Multipass: Last pass laced.	4–10 Small to large. Size on size.	Schedule 40	E6010	⅛ & 5/32	EP	522	15-J55	Advanced level
16-J13	45° L	Groove fillet	Horizontal fixed	Multipass: 1st pass stringer; others weaved.	4–10	Schedule 40	E6010	⅛ & 5/32	EP	522	15-J43–J55	Advanced level
16-J14	90° L	Groove fillet	Horizontal fixed	Multipass. 1st pass stringer; others weaved.	4–10	Schedule 40	E6010	⅛ & 5/32	EP	522	15-J43–J55	Advanced level
16-J15	Blunt pipe head	Groove fillet	Horizontal fixed	Multipass.	4–6	Schedule 40	E6010	⅛ & 5/32	EP	522	15-J43–J55	Advanced level
16-J16	Orange peel head	Groove fillet	Horizontal fixed	Multipass.	4–6	Schedule 40	E6010	⅛ & 5/32	EP	522	14-J33 15-J43–J55	Advanced level
16-J17	90° Y	Groove fillet	Fixed	Multipass.	4–6	Schedule 40	E6010	⅛ & 5/32	EP	522	14-J33 15-J43–J55	Advanced level

Notes:

At this point the student should hand-cut, bevel, and weld several pipe joints in all three positions.

The student should be able to pass bend tests of pipe coupons cut from pipe welded in the horizontal fixed (5G) and vertical fixed positions (2G).

The student should be able to produce acceptable pipe test pieces by using manual and machine oxyacetylene and plasma arc cutting equipment in order to comply with the AWS S.E.N.S.E. advanced level.

At this point the student should repeat those jobs on which additional practice is necessary, using pipe of different diameters and wall thicknesses. The student is now ready to take the official API or ASME Code Certification.

Test for Carbon Steel Pipe. Welding is to be in the horizontal fixed (5G) and the vertical fixed positions (2G).

Students who qualify on standard carbon steel pipe for SMAW may take additional training in SMAW of heavy, extra heavy, and alloy pipe and in GTAW and GMAW of alloy and aluminum pipe.

[1]E7010 electrodes should also be used for practice.

E6011 electrodes should be used with alternating current.

E7018, E9018, and E10018 electrodes should be used for low alloy steel pipe.

[2]For additional information follow AWS WPS ANSI/AWS B2.1-1-208 for carbon steel pipe.

[3]For additional information follow AWS WPS ANSI/AWS B2.1-1-201 for carbon steel pipe.

[4]For additional information follow AWS WPS AWS3-SMAW-1.

Table 17-11 Job Outline: Air Carbon Arc Cutting: Jobs 17-J6–J7

Job Number	Type of Job	Material[1]	Thickness (in.)	Carbon Diameter	Electrical Requirements DCEP		Minimum Air Pressure and Flow Rate	Text Reference	AWS S.E.N.S.E. Reference
					Amperes	Volts			
17-J6	Gouge U-groove	Carbon steel	¼	⁵⁄₁₆	400	42	60 p.s.i. @ 30 ft³/min	561	Entry level, advanced level[1]
17-J7	Weld removal	Carbon steel	¼	¼	350	42	60 p.s.i. @ 30 ft³/min	562	Entry level, advanced level[1]

[1]Practice is required on pipe and plate for the advanced level.

Table 19-15 Job Outline: Gas Tungsten Arc Welding Practice (Plate)

Job Number	Type of Joint	Type of Weld	Position[1]	Material Type[2]	Material Thickness (in.)	Tungsten Size (in.)	Current Type[3]	Current Amperes	Shielding Gas Type	Shielding Gas Flow (ft3/h)	Cup Size (in.)	Filler Rod Size (in.)	Text Reference	AWS S.E.N.S.E. Reference
19-J1	Flat plate	Beading	1	Aluminum	1/8	3/32	a.c.	90–140	Argon	12–20	3/8	3/32	631	Entry level
19-J2	Edge	Beading	1	Aluminum	1/8	3/32	a.c.	90–140	Argon	12–20	3/8	3/32	634	Entry level
19-J3	Outside corner	Fillet	1	Aluminum	1/8	3/32	a.c.	100–140	Argon	12–20	3/8	3/32	635	Entry & advanced levels
19-J4	Lap	Fillet	2	Aluminum	1/8	3/32–1/8	a.c.	90–140	Argon	12–20	3/8	3/32	636	Entry & advanced levels
19-J5	T	Fillet	2	Aluminum	1/8	3/32–1/8	a.c.	90–150	Argon	12–20	5/16	3/32	637	Entry & advanced levels
19-J6	Square butt	Groove	1	Aluminum	1/8	3/32	a.c.	80–130	Argon	12–20	3/8	3/32	639	Entry & advanced levels
19-J7	Single V butt	Groove	1	Aluminum	1/8	1/16	a.c.	190–280	Argon	22–28	1/2	3/32–1/8	639	Entry & advanced levels
19-JQT1[4]	Butt & T	Fillet & groove	1 & 2	Aluminum	18–10 gauge	1/8 max.	a.c.	40–125	Argon	20–40	1/4–3/8	3/32–1/8	Fig. 19-44	Entry level
19-J8	Flat plate	Beading	1	Mild steel	1/8	1/16–3/32	DCEN	90–120	Argon	10–14	1/4–3/8	3/32	641	Entry level
19-J9	Edge	Beading	1	Mild steel	1/8	1/16–3/32	DCEN	90–120	Argon	10–14	1/4–3/8	3/32	641	Entry level
19-J10	Outside corner	Fillet	1	Mild steel	1/8	1/16–3/32	DCEN	90–120	Argon	10–14	1/4–3/8	3/32	641	Entry level
19-J11	Lap	Fillet	2	Mild steel	1/8	1/16–3/32	DCEN	100–130	Argon	10–14	1/4–3/8	3/32	641	Entry level
19-J12	T	Fillet	2	Mild steel	1/8	1/16–3/32	DCEN	100–130	Argon	10–14	1/4–3/8	3/32	641	Entry level
19-JQT2[5]	Butt & T	Fillet & groove	1, 2, 3, & 4	Mild steel	18–10 gauge	1/8 max.	DCEN	45–130	Argon	15–25	1/4–5/8	1/16–3/32	Fig. 19-45	Entry level
19-J13	Flat plate	Beading	1	Stainless steel	1/8	1/16–3/32	DCEN	80–130	Argon	10–12	1/4–3/8	3/32	641	Entry level
19-J14	Outside corner	Fillet	2	Stainless steel	1/8	1/16–3/32	DCEN	80–130	Argon	10–12	1/4–3/8	3/32	641	Entry & advanced levels
19-J15	Lap	Fillet	2	Stainless steel	1/8	1/16–3/32	DCEN	80–140	Argon	10–12	1/4–3/8	3/32	641	Entry & advanced levels
19-J16	T	Fillet	2	Stainless steel	1/8	1/16–3/32	DCEN	80–140	Argon	10–12	1/4–3/8	3/32	641	Entry & advanced levels
19-J17	Square butt	Groove	1	Stainless steel	1/8	1/16–3/32	DCEN	70–120	Argon	10–12	1/4–3/8	3/32	641	Entry & advanced levels
19-J18	Single V butt	Groove	1	Stainless steel	1/8	3/32	DCEN	90–150	Argon	12–15	3/16–1/2	3/32–1/8	641	Entry & advanced levels
19-J19	Single V butt	Groove	1	Stainless steel	1/4	1/8	DCEN	110–180	Argon	12–15	1/2	1/16–3/32	641	Entry & advanced levels
19-JQT3[6]	Butt & T	Fillet & groove	1, 2, & 3	Stainless steel	18–10 gauge	1/8 max.	DCEN	35–130	Argon	15–25	1/4–3/8	1/16–3/32	Fig. 19-46	Entry level

Notes:
- Also practice in the vertical and overhead positions.
- For vertical and overhead welding, reduce the amperage by 10 to 20 amperes.
- When two rod and tungsten sizes are listed, the smaller is for vertical and overhead position welding.
- Ceramic or glass cups should be used for currents to 250 amperes.
- Water-cooled cups should be used for currents above 250 amperes.
- The gas flow should be set at the maximum rate for vertical and overhead welding.
- The type of filler rod must always match the type of base metal being welded.
- This table is typical. Tungsten electrode size, filler rod size, and welding current will vary with the welding situation and the welder's skill.
- Visual examination will be done in accordance with the requirements of AWS QC10: 2008, Table 3.

[1]1 = flat, 2 = horizontal, 3 = vertical, 4 = overhead.
[2]Also practice on nickel, copper, magnesium, and titanium to comply with the AWS S.E.N.S.E. expert level.
[3]a.c. with stabilization.
[4]For additional information follow AWS Standard Welding Procedure Specification (WPS) B2.1-015 for aluminum.
[5]Also practice in the vertical and overhead positions prior to doing this qualification test. For additional information follow AWS Standard Welding Procedure Specification (WPS) B2.1-008 for carbon steel.
[6]Also practice in the vertical and overhead positions prior to doing this qualification test. For additional information follow AWS Standard Welding Procedure specification (WPS) B2.1-009 for stainless steel.

Table 20-2 Job Outline: Gas Tungsten Arc Welding Practice (Pipe)

Recommended Job Order[1]	Number in Text	Type of Joint	Type of Weld	Position[2]	Welding Technique	Pipe Specifications		
						Material	Diameter (in.)	Wall Thickness
1st	20-J1	NA	Beading	1 (pipe rotated)	Stringer-up, weave-up	Steel	4–6	¼ in.
2nd	20-J2	Butt	V-groove	1 (pipe rotated)	1 stringer-up, 1 weave-up	Steel	4–6	¼ in.
3rd	20-J8	Butt	V-groove	2	5 stringers	Steel	4–6	¼ in.
4th	20-J12	Butt	V-groove	5	1 stringer-up, 1 weave-up	Steel	4–6	¼ in.
Test	20-JQT1[7]	Butt & T	Groove & fillet	5	Stringer or weave	Steel	3	18–10 gauge
Test	20-JQT2[8]	Butt	Sq. or V-groove	2 & 5	Stringer or weave	Steel	1–2⅞	18–10 gauge
Test	20-JQT3[9]	Butt	Sq. or V-groove	6	Stringer or weave	Steel	2½	0.05–0.120 in.
5th	20-J3	NA	Beading	1 (pipe rotated)	Stringer-up, weave-up	Al.	4–6	¼ in.
6th	20-J4	Butt	V-groove	1 (pipe rotated)	1 stringer-up, 1 weave-up	Al.	4–6	¼ in.
7th	20-J9	Butt	V-groove	2	5 stringers	Al.	4–6	¼ in.
8th	20-J13	Butt	V-groove	5	1 stringer-up, 2 weave-up	Al.	4–6	¼ in.
9th	20-J14	Butt	V-groove	5	1 semiweave	Al.	2–3	⅛ in.
10th	20-J15	Butt	V-groove	5	1 semiweave	Al.	2–3	⅛ in.
11th	20-J10	Butt	V-groove	2	1 semiweave	Al.	2–3	⅛ in.
Test	20-JQT4[7]	Butt & T	Groove & fillet	5	Stringer or weave	Al.	3	18–10 gauge
Test	20-JQT5[8]	Butt	Sq. or V-groove	2 & 5	Stringer or weave	Al.	1–2⅞	18–10 gauge
Test	20-JQT6[9]	Butt	Sq. or V-groove	6	Stringer or weave	Al.	2½	0.05–0.120 in.
12th	20-J5	NA	Beading	1 (pipe rotated)	Stringer-up, weave-up	S. Stl	4–6	¼ in.
13th	20-J6	Butt	V-groove	1 (pipe rotated)	1 stringer-up, 1 weave-up	S. Stl	4–6	¼ in.
14th	20-J11	Butt	V-groove	2	5 stringers	S. Stl	4–6	¼ in.
15th	20-J16	Butt	V-groove	5	1 stringer-up, 2 weave-up	S. Stl	4–6	¼ in.
16th	20-J7	Butt	Sq.-groove	1	1 semiweave	S. Stl	2–3	⅛ in.
17th	20-J17	Butt	Sq.-groove	5	1 semiweave	S. Stl	2–3	⅛ in.
Test	20-JQT7[7]	Butt & T	Groove & fillet	5	Stringer or weave	S. Stl	3	18–10 gauge
Test	20-JQT8[8]	Butt	Sq. or V-groove	2 & 5	Stringer or weave	S. Stl	1–2⅞	18–10 gauge
Test	20-JQT9[9]	Butt	Sq. or V-groove	6	Stringer or weave	S. Stl	2½	0.05–0.120 in.

[1]It is recommended that the student do the jobs in this order. In the text, the jobs are grouped according to the type of operation to avoid repetition. The tests will be administered per your instructor's instructions.

[2]1 = flat, 2 = horizontal, 5 = multiple, 6 = multiple axis 45°.

[3]Ceramic or glass cups should be used for currents up to 250 A. Water-cooled cups should be used for currents above 250 A.

[4]The gas flow should be set for the maximum rate for the vertical and overhead positions. Gas backing is optional for steel and aluminum. It is required for stainless steel.

[5]The type of filler rod must always match the type of base metal being welded.

[6]NA = not appropriate.

Tungsten Electrode Size (in.)	Current		Shielding gas		Cup Size (in.)	Filler Rod Size[5] (in.)	Text Reference	AWS S.E.N.S.E. Reference[6]
	Type	Amperes[3]	Gas Type	Gas Flow (ft3/h)[4]				
3/32	DCEN	140–160	Argon	8–10	3/8	3/32–1/8	654	NA
3/32	DCEN	120–140 140–160	Argon	8–10	3/8	3/32–1/8	654	Entry level
3/32	DCEN	120–160	Argon	8–10	3/8	3/32–1/8	659	Advanced level
3/32	DCEN	120–140	Argon	8–10	3/8	3/32–1/8	661	Advanced level
1/8 max.	DCEN	45–130	Argon	15–25	1/4–5/8	1/16–3/32	Fig. 20-49	Advanced level
1/8 max.	DCEN	45–130	Argon	15–25	1/4–5/8	1/16–3/32	Fig. 20-50	Advanced level
1/8 max.	DCEN	45–130	Argon	15–25	1/4–5/8	1/16–3/32	Fig. 20-51	Expert level
1/8	A.C.	150–280	Argon	22–28	1/2	3/32–1/8	654	NA
1/8	A.C.	150–280 160–300	Argon	22–28	1/2	3/32–1/8	654	Entry level
1/8	A.C.	150–200 160–210	Argon	25–30	1/2	3/32–1/8	659	Advanced level
1/8	A.C.	120–200 140–220	Argon	25–30	1/2	3/32–1/8	661	Advanced level
3/32	A.C.	120–145	Argon	15–20	3/8	3/32–1/8	661	Advanced level
3/32	A.C.	100–120	Argon	12–22	3/8	3/32–1/8	661	Advanced level
3/32	A.C.	100–120	Argon	15–20	3/8	3/32–1/8	659	Advanced level
1/8 max.	A.C.	40–125	Argon	20–40	1/4–5/8	3/32–1/8	Fig. 20-49	Advanced level
1/8 max.	A.C.	40–125	Argon	20–40	1/4–5/8	3/32–1/8	Fig. 20-52	Advanced level
1/8 max.	A.C.	40–125	Argon	20–40	1/4–5/8	3/32–1/8	Fig. 20-51	Expert level
3/32	DCEN	120–160	Argon	12–15	1/2	3/32–1/8	654	NA
3/32	DCEN	110–150 110–160	Argon	12–15	1/2	3/32–1/8	654	Entry level
3/32	DCEN	110–150 110–160	Argon	12–15	1/2	3/32–1/8	659	Advanced level
3/32	DCEN	110–150 110–160	Argon	14–18	1/2	3/32–1/8	661	Advanced level
1/16 to 3/32	DCEN	60–120	Argon	12–15	3/32–1/8	3/32–1/8	654	Advanced level
	DCEN	80–140	Argon	12–15	3/32–1/8	3/32–1/8	661	Advanced level
1/8 max.	DCEN	35–130	Argon	15–25	1/4–5/8	1/16–3/32	Fig. 20-49	Advanced level
1/8 max.	DCEN	35–130	Argon	15–25	1/4–5/8	1/16–3/32	Fig. 20-53	Advanced level
1/8 max.	DCEN	35–130	Argon	15–25	1/4–5/8	1/16–3/32	Fig. 20-51	Expert level

[7]For additional information follow AWS Standard Welding Procedure Specification (WPS) B2.1.008 for carbon steel, AWS2-1-GTAW or AWS2-1.1-GTAW for aluminum, and (WPS) B2.1.009 for stainless steel.

[8]For additional information follow AWS Standard Welding Procedure Specification (WPS) B2.1.008 for carbon steel.

[9]For additional information follow AWS Standard Welding Procedure Specification (WPS) B2.1.008 for carbon steel; AWS3-GTAW-1, AWS3-GTAW-2, AWS2-1-GTAW, or AWS2-1.1-GTAW for aluminum; and AWS B2.1.009 and AWS B2.1.010 for stainless steel.

Note: The conditions given here are basic. They vary with the job situation, the results desired, and the skill of the welder. It is also generally acknowledged if you are capable of welding a more difficult weld and joint like a groove weld on a butt joint that you are capable of making easier fillet welds on lap joints and T-joints.

Table 22-8 Job Outline: Gas Metal Arc Welding Practice with Solid Core Wire (Plate)

Recommended Job Order[1]	Number in Text	Material: Type	Material: Thickness	Type of Weld	Type of Joint	Weld Position	No. of Passes	Electrode[2]: Type AWS	Electrode[2]: Size (in.)	Shielding Gas	Gas Flow (ft³/h)	Welding Current[3] DCEP: Arc (V)	Welding Current[3] DCEP: Amperes	Welding Current[3] DCEP: Wire-Feed Speed[4] (WFS) (in./min)	Text Reference	AWS S.E.N.S.E. Reference[5]
1st	22-J1	Carbon steel	⅛	Groove	Square butt	Flat (1G)	2	E70S-3 ER70S-6	0.035	Carbon dioxide	20–25	19–21	110–160	170–340	751	Entry level
2nd	22-J3	Carbon steel	3/16	Fillet	Lap	Flat (1F)	1	E70S-3 ER70S-6	0.035	Carbon dioxide	20–25	19–21	110–170	170–360	752	Entry level
3rd	22-J4	Carbon steel	3/16	Fillet	Lap	Vertical down- hill (3F)	1	E70S-3 ER70S-6	0.035	Carbon dioxide	20–25	19–21	120–160	190–340	752	Entry level
4th	22-J5	Carbon steel	3/16	Fillet	Lap	Overhead (4F)	1	E70S-3 ER70S-6	0.035	Carbon dioxide	20–25	19–21	120–160	190–340	752	Entry level
5th	22-J6	Carbon steel	⅛	Fillet	T	Horizontal (2F)	1	E70S-3 ER70S-6	0.035	Argon 75% carbon dioxide 25%	20–25	19–21	110–170	170–360	752	Entry level
6th	22-J7	Carbon steel	¼	Fillet	T	Horizontal (2F)	1	ER70S-3	0.045	Argon 98% Oxygen 2%	40–50	24–32	200–375	225–410	752	Advanced level
7th	22-J8	Carbon steel	⅜	Fillet	T	Vertical uphill (3F)	3	ER70S-6	0.035	Carbon dioxide	20–25	21–23	150–160	300–340	752	Entry level
8th	22-J9	Carbon steel	⅜	Fillet	T	Horizontal (2F)	3	ER70S-3	1/16	Argon 95% Oxygen 5%	40–50	26–33	275–400	200–280	752	Advanced level
9th	22-J10	Carbon steel	⅜	Fillet	T	Overhead (4F)	6	ER70S-6	0.035	Carbon dioxide	20–25	21–23	160–180	340–380	752	Entry level
10th	22-J2	Carbon steel	½	Groove	V-butt 60°	Vertical downhill root (3G) Vertical uphill (3G)	2 2	ER70S-6	0.035	Carbon dioxide	20–25 20–25	21–23 21–23	160–190 160–180	340–400 340–380	751 751	Entry level
22-J1–J10	22-JQT1[5]	Carbon steel	⅜	Fillet and groove	Butt, lap, T-, corner	2F 3F 4F 2G 3G 4G	As required to get weld size	ER70S-3	0.035	Argon 75% Carbon dioxide 25%	20–25	19–21	160–180	340–380	753	Entry level
22-J1–J10	22-JQT2[5]	Carbon steel	⅜	Fillet and groove	Butt, T	2F 1G	As required to get weld size	ER70S-3	0.045	Argon 98% Oxygen 2%	40–50	24–32	200–275	225–410	754	Entry level
22-J1–J10	22-JQT3[5]	Carbon steel	1	Groove	Butt	1G	As required to get weld size	ER70S-3	0.045	Argon 98% Oxygen 2%	40–50	24–32	200–275	225–410	757	Advanced level
11th	22-J11	Aluminum	⅛	Groove	Square butt	Flat (1G)	2	ER1100	3/64	Argon	30–35	20–24	150–190	280–400	760	Advanced level
12th	22-J13	Aluminum	⅛	Fillet	Lap	Horizontal (2F)	1	ER1100	3/64	Argon	30–35	20–24	150–190	280–400	761	Advanced level
13th	22-J14	Aluminum	3/16	Fillet	T	Horizontal (2F)	3	ER1100	3/64	Argon	30–35	21–25	160–200	290–430	761	Advanced level
14th	22-J15	Aluminum	¼	Fillet	T	Vertical uphill (3F)	2	ER4043	3/64	Argon	30–35	21–25	160–190	290–400	761	Advanced level
15th	22-J16	Aluminum	¼	Fillet	T	Overhead (4F)	3	ER4043	3/64	Argon	30–35	21–25	170–190	320–400	761	Advanced level

16th	22-J12	Aluminum	⅜	Groove	V-butt 60% backup	Vertical uphill (3G)	4	ER4043	1/16	Argon 75% Helium 25%	30–40	22–26	200–250	260–330	760	Advanced level
22-J11–J16	22- JQT45	Aluminum	⅜	Groove	Butt	3G 4G	As required to get weld size	ER4043	3/64	Argon	30–35	21–25	160–190	320–400	763	Advanced level
17th	22-J17	Stainless steel	⅛	Beading	Plate	Flat (1C)	Cover plate	ER308	0.035	Helium 90% Argon 7½% Carbon dioxide 2½%	22–24	24–26	130–160	160–280	766	Advanced level
18th	22-J18	Stainless steel	¼	Beading	Plate	Flat (1C)	Cover plate	ER308	0.035	Argon 98% Oxygen 2%	25–35	24–27	130–160	160–280	766	Advanced level
19th	22-J19	Stainless steel	⅛	Fillet	Lap	Horizontal	1	ER308	0.035	Helium 90% Argon 7½% Carbon dioxide 2½%	22–24	24–26	140–160	175–280	766	Advanced level
20th	22-J20	Stainless steel	⅛	Fillet	T	Horizontal	1	ER308	0.035	Helium 90% Argon 7½% Carbon dioxide 2½%	22–24	24–26	150–180	190–290	766	Advanced level
21st	22-J21	Stainless steel	¼	Fillet	T	Horizontal	6	ER308	0.035	Argon 95% Oxygen 5%	25–35	24–27	180–240	290–390	766	Advanced level
22nd	22-J22	Stainless steel	¼	Fillet	T	Vertical uphill (3F)	6	ER308	0.035	Argon 98% Oxygen 2%	25–35	20–26	140–200	175–320	766	Advanced level
23rd	22-J23	Stainless steel	⅜	Groove	V-butt 60° backup	Vertical uphill (3G)	1 down 5 up	ER308	0.035	Argon 98% Oxygen 2%	25–35	20–28	140–200	175–320	766	Advanced level

Note: The conditions indicated here are basic. They will vary with the job situation, the results desired, and the skill of the welder.

[1]It is recommended that the student do the jobs in this order. In the text, the jobs are grouped according to the type of operation to avoid repetition.

[2]On all carbon steel work, use metal cored wire to practice (E70C-1C or E70C-1M, depending on the shielding gas being used). You will need to increase the wire-feed speed or go to the next sized electrode diameter to compensate for the higher current density.

[3]Pulse spray arcs should be practiced on all jobs. Use the equipment manufacturer's recommended parameter setting. Meet or exceed the wire-feed speed for the other modes of transfer.

[4]WFS = wire-feed speed.

[5]Refer to AWS S.E.N.S.E. documents QC10:2008, QC11-96, and QC12-96 for additional information.

Table 23-10 Job Outline: Flux Cored Arc Welding Practice with Gas Shielded Electrodes (Plate)

Recommended Job Order	Material	Plate Thickness (in.)	Joint	Weld	Weld Position	No. of Passes	Electrode[1] Type AWS	Size (in.)	Shielding Gas[2] Type	Gas Flow (ft³/h)	Welding Current Volts	Amperes	Wire-Feed Speed (WFS) (in./min)	Electrode Extension (in.)	AWS S.E.N.S.E. Reference
1st	Steel	¼	Flat plate	Surfacing	Flat (1C)	Cover plate	E71T-1	¹⁄₁₆	Carbon dioxide	30–40	25–30	310–365	275–375	1	Not applicable
2nd	Steel	¼	T	Fillet	Flat (1F)	2	E71T-1	¹⁄₁₆	Carbon dioxide	30–40	25–30	310–365	275–375	1	Entry level
3rd	Steel	¼	Lap	Fillet	Horizontal (2F)	1	E71T-1	¹⁄₁₆	Carbon dioxide	30–40	23–30	310–365	275–375	1	Entry level
4th	Steel	⅜	T	Fillet	Horizontal (2F)	6	E71T-1	¹⁄₁₆	Carbon dioxide	35–40	23–30	310–365	275–375	1	Entry level
5th	Steel	³⁄₁₆	Butt	Square-groove with backing	Flat (1G)	1	E71T-1	¹⁄₁₆	Carbon dioxide	30–40	23–30	310–365	275–375	1	Entry level
6th	Steel	⅜	Butt	V-groove with backing	Flat (1G)	2	E71T-1	¹⁄₁₆	Carbon dioxide	35–40	23–30	310–365	275–375	1	Entry level
7th	Steel	⅜	T	Fillet	Vertical (3F)	2	E71T-1	0.045	Carbon dioxide	35–40	25–27	180–220	275–340	¾	Entry level
8th	Steel	½	Butt	V-groove with backing	Vertical (3G)	3	E71T-1	0.045	Carbon dioxide	35–40	25–27	180–220	275–340	¾	Entry level
9th	Steel	½	Butt	V-groove with backing	2G	6	E71T-1	0.045	Carbon dioxide	35–45	26–30	220–275	340–500	¾	Entry level
10th	Steel	½	T	Fillet	4F	3	E71T-1	0.045	Carbon dioxide	35–45	25–27	180–220	275–340	¾	Entry level
11th	Steel	½	Butt	V-groove with backing	4G	6	E71T-1	0.045	Carbon dioxide	35–45	27–28	190–220	300–340	¾	Entry level
23-JQT1[3]	Steel	⅜	Butt, lap, T	Fillet and groove	2F 3F 4F 2G 3G	As required to attain weld size	E71T-1	0.045	Carbon dioxide	35–45	25–30	180–275	275–500	¾	Entry level qualification test

Note: The conditions given here are basic and will vary with the job situation, the results desired, and the skill of the welder.

[1] Other FCAW-G electrode types and sizes may be substituted. The specific manufacturer's parameter recommendations should be used. E70T-1 electrodes can be used for the flat (1) and horizontal (2) positions.

[2] Use of 75 to 80% Argon, balance CO₂ can be used with the E71T-1M electrodes; for less smoke, spatter, and a very smooth stable arc. Penetration will be reduced from 100% CO₂. Better weld pool control will be provided for out-of-position welding.

[3] Refer to AWS S.E.N.S.E. documents QC10:2008, and QC11-96 for additional information.

Table 23-11 Job Outline: Flux Cored Arc Welding Practice with Self-Shielded Electrodes

							Electrode[1]		Welding Current DCEN				
Recommended Job Order	Material	Plate Thickness	Joint	Weld	Weld Position	No. of Passes	Type	Size (in.)	Arc Volts	Amperes	Wire-Feed Speed (in./min)	Electrode Extension (in.)	AWS S.E.N.S.E. Reference
1	Steel	3/16 in.	Flat plate	Surfacing	1C	Cover plate	E71T-11	0.068	18–20	190–270	75–130	3/4	Entry level
2	Steel	3/16 in.	Butt	Square groove with backing	1G	1	E71T	0.068	15–18	125–190	40–75	5/8	Entry level
3	Steel	3/16 in.	T	Fillet	2F	1	E71T	0.068	18–20	190–270	75–130	5/8	Entry level
4	Steel	10 gauge	T	Fillet	2F	1	E71T	0.068	15–18	125–190	40–75	3/4	Entry level
5	Steel	1/4 in.	T	Fillet	3F	1	E71T	0.045	15–18	140–160	90–110	3/8	Entry level
6	Steel	3/16 in.	Lap	Fillet	1F	1	E71T	0.068	18–20	190–270	75–130	3/4	Entry level
7	Steel	10 gauge	Lap	Fillet	1F	1	E71T	0.068	18–18	125–190	40–75	3/4	Entry level
8	Steel	1/4 in.	T	Fillet	4F	1	E71T	0.045	16–17	140–160	90–110	3/8	Entry level
9	Steel	3/8 in.	Butt	V-groove with backing	1G	2	E71T	0.068	20–23	270–300	130–175	3/4	Entry level
10	Steel	1/2 in.	Butt	V-groove with backing	2G	1 2–3 4–6	E71T	0.045	15–16 16–17 17–18	120–140 140–160 160–170	70–90 90–110 110–130	3/8	Entry level
11	Steel	1/2 in.	Butt	V-groove with backing	3G	1–3	E71T	0.045	17–18	160–170	110–130	3/8	Entry level
12	Steel	1/2 in.	Butt	V-groove with backing	4G	1 2–3 4–6	E71T	0.045	15–16 16–17 17–18	124–140 140–160 160–170	70–90 90–110 110–130	3/8	Entry level
23-JQT2[2]	Steel	3/8 in.	Butt, lap, T	Grooves and fillets	2F 3F 4F 2G 3G	As required to attain weld size	E71T	0.045	16–18	140–170	90–130	3/8	Entry level

[1]Other FCAW-S electrode types and sizes may be substituted. Always use the specific manufacturer's parameter recommendations. Because of the unique nature of the self-shielded electrode the Lincoln Innershield NR-211MP was used for the development of this table.

[2]Refer to AWS S.E.N.S.E. documents QC10-2008 and QC11-96 for additional information.

Table 23-12 Job Outline: Semiautomatic Application of the Submerged Arc Welding Process

Recommended Job Order	Material	Plate Thickness (in.)	Joint	Weld	Weld Position	No. of Passes	Electrode¹/flux Type	Size (in.)	Welding Current DCEN Arc Volts	Amperes	Electrode Extension (in.)	AWS S.E.N.S.E. Reference
1	Steel	⅜	T	Fillet	1F	1	F6A2-E1M12	¹⁄₁₆	37	400	1–2	Not applicable

Table 24-2 Job Outline: Gas Metal Arc Welding Practice with Solid Wire (Pipe)

Job No.	Type of Joint	Type of Weld	Welding Position	Welding Technique	Pipe Specifications — Material	Diameter (in.)	Weight Schedule	Wall Thickness (in.)	Electrode Specifications[1] Type	Size (in.)	Welding Current DCEP — Arc Volts	Amperes	Wire-Feed Speed	Shielding Gas[2]	Gas Flow (ft³/h)	Text Reference	AWS S.E.N.S.E. Reference[3]
24-J1	Flat surface	Surfacing	1C	Stringer downhill / Weaved downhill	Carbon steel	4–8	40	$\frac{1}{4}$–$\frac{5}{16}$	ER70S-6	0.035	18–21	150–160 / 130–150	250–290 / 220–250	Carbon dioxide	15–25	817	NA
24-J2	Flat surface	Surfacing	1C	Stringer uphill / Weaved uphill	Carbon steel	4–8	80	$\frac{5}{16}$–$\frac{1}{2}$	ER70S-6	0.035	19–23	130–160	220–290	Carbon dioxide	15–20	817	NA
24-J3	Butt	V-groove	1G	1 Stringer—downhill / 2 Weaved—downhill	Carbon steel	6	40	$\frac{3}{16}$	ER70S-6	0.035	18–21	130–140 / 140–160	220–235 / 235–290	Carbon dioxide	15–20	819	NA
24-J4	Butt	V-groove	2G	5 Stringer	Carbon steel	6	40	$\frac{3}{16}$	ER70S-6	0.035	18–21	1–130 to 140 / 4–140 to 160	220–235 / 235–290	Carbon dioxide	15–25	821	Advanced level
24-J5	Butt	V-groove	2G	7 Stringer	Carbon steel	8	80	$\frac{1}{2}$	ER70S-6	0.035	19–23	1–140 to 150 / 6–150 to 170	235–250 / 250–330	Carbon dioxide	15–25	821	Advanced level
24-J6	Butt	V-groove	5G	1 Stringer—downhill / 3 Weaved—downhill	Carbon steel	8	40	$\frac{3}{16}$	ER70S-6	0.035	19–21	120–130 / 130–150	210–220 / 220–250	Carbon dioxide	15–20	823	Advanced level
24-J7	Butt	V-groove	5G	1 Stringer—downhill / 2 Weaved—uphill	Carbon steel	8	40	$\frac{3}{16}$	ER70S-6	0.035	19–21 / 19–23	120–130 / 110–120	210–220 / 180–210	Carbon dioxide	15–20	827	Advanced level
24-J8	Butt	V-groove	5G	1 Stringer—downhill / 2 Weaved—uphill	Carbon steel	8	80	$\frac{1}{2}$	ER70S-6	0.035	19–21 / 19–24	120–130 / 110–125	210–220 / 180–215	Carbon dioxide	15–20	827	Advanced level
24-J9	90° branch	V-groove fillet	Header horizontal fixed; branch vertical fixed top (2F, 2G)	3 Stringer	Carbon steel	6–8	40	$\frac{3}{16}$	ER70S-6	0.035	18–21	130–140	220–235	Carbon dioxide	15–20	829	Advanced level

(Continued)

Table 24-2 (Concluded)

24-J10	90° branch	V-groove, fillet	Header vertical fixed; branch horizontal fixed (5F, 5G)	1 Stringer—downhill / 2 Weaved uphill	Carbon steel	6–8	40	3/16	ER70S-6	0.035	18–21	120–130 / 110–120	210–220 / 180–210	Carbon dioxide	15–20	829	Advanced level
24-J11	90° branch	V-groove, fillet	Header horizontal fixed; branch vertical fixed bottom (5F, 5G)	3 Stringer	Carbon steel	6–8	40	3/16	ER70S-6	0.035	18–21	120–130	210–220	Carbon dioxide	15–20	829	Advanced level
24-J12	45° branch	V-groove, fillet	Header horizontal fixed; branch top	3 Stringer	Carbon steel	6–8	40	3/16	ER70S-6	0.035	18–21	130–140	220–235	Carbon dioxide	15–20	829	Advanced level
24-J13	45° branch	V-groove, fillet	Header vertical fixed; branch side	1 Stringer—downhill / 2 Weaved uphill	Carbon steel	6–8	40	3/16	ER70S-6	0.035	18–21 / 18–23	120–130 / 110–120	210–220 / 180–210	Carbon dioxide	15–20	829	Advanced level
24-J14	45° branch	V-groove, fillet	Headers horizontal fixed; branch bottom	5 Stringer	Carbon steel	6–8	40	3/16–1/2	ER70S-6	0.035	18–23	130–140	220–235	Carbon dioxide	15–20	829	Advanced level
25-J15	Butt	V-groove	Horizontal roll (1G)	1 Stringer / 1 Weaved	Aluminum	5–6	80	1/4	ER4043	3/64	20–24	180–190 / 190–210	365–390 / 390–426	Argon	30–40	831	NA
24-JQT¹	T and lap	Groove, fillet	2F, 5F, 5G	Stringer and weave	Carbon steel	4–6	40	1/4–3/16	ER70S-6	0.035	18–23	110–160	180–290	Carbon dioxide	15–25	829	Advanced level
24-JQT²	Butt	Groove	6G	Stringer and weave	Carbon steel	2½–6	40	0.203–3/16	ER70S-6	0.035	19–21	120–150	210–250	Carbon dioxide	15–25	831	NA
24-JQT³	Butt	Groove	6G	Stringer and weave	Aluminum	2½–6	40	0.203–3/16	ER4043	3/64	(4)	(4)	(4)	Argon	30–40	832	NA

[1] Electrode type ER 80S-D2 may also be used with excellent results.

[2] Argon-rich gas mixtures may also be used. GMAW-P may also be used if equipment is available. Various root pass techniques can also be used, such as GTAW and GMAW-S. GMAW-P is very difficult unless the joint is accurately fit and welding is done in the 1G position.

[3] NA = not applicable.

[4] Use a synergic pulse parameter as specified by your specific equipment manufacturer for 3/64-in. wire.

Note: The conditions indicated here are basic and will vary with the job situation, the results desired, and the skill of the welder. For additional practice the FCAW process can be substituted. Stainless-steel practice can be done using appropriate filler metal and shielding gases. Stainless-steel pipe is expensive and limited practice can be done using carbon steel pipe with stainless electrode and appropriate shielding gas.